Experimental Microbial Ecology

Experimental Microbial Ecology

SHELDON AARONSON

Department of Biology
Queens College
City University of New York
Flushing, New York

ACADEMIC PRESS New York and London 1970

ACADEMIC PRESS, INC.
111 Fifth Avenue, New York, New York 10003

United Kingdom Edition published by
ACADEMIC PRESS, INC. (LONDON) LTD.
24/28 Oval Road, London NW1

LIBRARY OF CONGRESS CATALOG CARD NUMBER: 70-91422
Second Printing, 1972

PRINTED IN THE UNITED STATES OF AMERICA

This book is affectionately dedicated to Dr. S. H. Hutner who has kindled and fueled an interest in microorganisms for so many, and produced a generation of microbiologists without benefit of degree-granting license

Contents

Preface

"In view of our present ignorance of ecological relations, we now cannot isolate more than a very small fraction of the immense variety of microbial types at will, and even if this were possible, such isolation would involve the expenditure of considerable time and effort."

"In communities that comprise various kinds of microbes, the components are also subject to competitive interactions which are of the utmost importance as determinants of evolution and natural selection. Our knowledge in this area is still quite rudimentary, however.—We still need more comprehensive studies on the effects of environmental factors on the composition of natural communities of microorganisms."

C. B. van Niel
Quart. Rev. Biol. **41**
pp. 105–112, 1966

As these quotations from van Niel indicate, there is much that has to be done in microbial ecology. This book is written in the hope that it will encourage and help others to overcome the gaps in our knowledge of the microbial flora and fauna and their environment. Hopefully this knowledge may also help to mitigate some of the man-made disasters in our environment such as water pollution with detergents, sewage, industrial wastes, and petroleum (Torrey Canyon disaster) and soil and water pollution with insecticides and herbicides.

I

Brief Survey

Microorganisms were seen for the first time in the seventeenth century by Leeuwenhoek, but not until the second half of the nineteenth century did man grasp how these tiny organisms played essential roles in nature and man's affairs.

Microorganisms were first associated with human, animal, plant diseases, and fermentations during the mid-nineteenth century thanks to Pasteur, Koch, Ehrlich, and many others. The pioneering work of Winogradsky and Beijerinck at the end of that century alerted scientists to the vital role of microorganisms in the nitrogen and sulfur cycles in soil and water. The realization soon followed that microorganisms are at the base of the food chain, especially those microorganisms found in water. Beijerinck's followers, notably Kluyver and his student van Niel, as well as many others then exploited microorganisms for studying the chemical activities of the cell (metabolism) and the relationship of these chemical activities to heredity, and vice versa. The stage was set for the unity of biochemistry and the revolution sweeping biological fundamentals toward intimacy with chemistry and physics.

The microorganisms discussed here include bacteria, higher fungi, algae, and protozoa. A more detailed description of the major categories of microorganisms is to be found in the references at the end of this book.

Microorganisms, notably bacteria and fungi, are primarily responsible for recycling the bodies of dead animals and plants into the CO_2, NH_3, and H_2O necessary for photosynthetic bacteria, algae, and higher plants to convert to their organic compounds. So this mineralization-synthesis cycle repeats itself as it courses through growth, death, and decay. Fortunately there is a seldom-broken rule: *if an organic compound is made by one form of life, another form of life uses it for carbon and energy* (see Fig. I-1).

As with the carbon cycle, microorganisms are also responsible for maintaining the fertility of soil and water by converting the atmospheric nitrogen and the nitrogen in organic compounds into nitrate and ammonia—compounds available to higher plants. Nitrogen-fixing bacteria and blue-green algae are the major converters of atmospheric nitrogen into compounds useful for plant growth (Fig. I-2).

Aerobic photosynthetic microorganisms, particularly those in water, provide most of the oxygen used by other organisms by splitting water during photosynthesis.

Bacteria are also responsible for converting large quantities of inorganic sulfur compounds from their reduced forms, H_2S or sulfur, to sulfite or sulfate, and vice versa. Bacteria can also concentrate iron salts in the soil by their chemical activities—indeed, the Mesabi iron range may be essentially a graveyard of iron-impregnated bacterial carcasses. Chemical activities by bacteria such as these are respon-

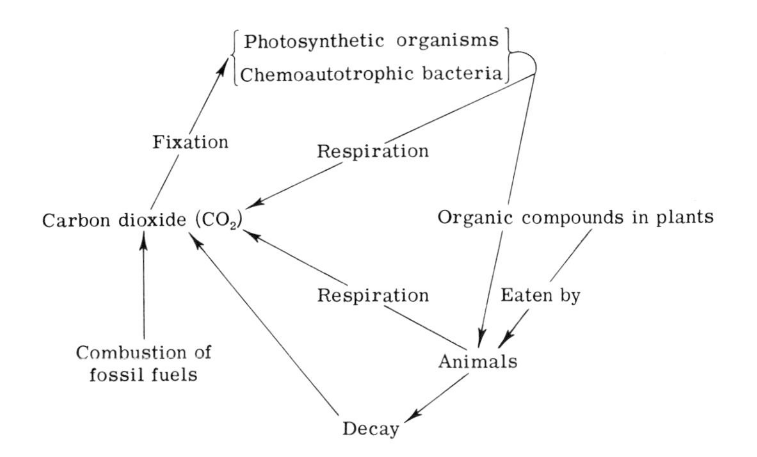

FIG. I-1. Carbon cycle.

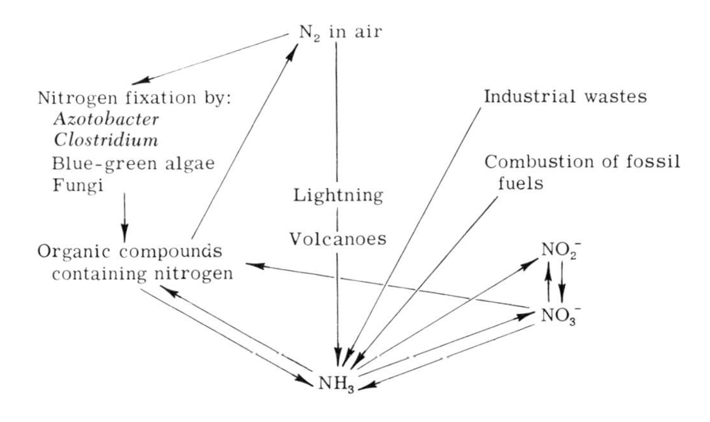

FIG. I-2. Nitrogen cycle.

sible for the unusually rich sulfur and iron deposits found in various parts of the world.

Microorganisms live and multiply in all the environments found on the earth: in soil; fresh, brackish, and marine water, at all depths; on ice and snow; on mountain tops; and in the air, where they may not multiply.

Soil

Soil is composed of mineral or organic matter, water, air, and organisms, the concentration of each varying with the type of soil and its source. Soils also vary in color, depth, and pH. Soils differ in texture; some soils contain large quantities of clay, others large quantities of sand or silt, while still other soils are rich in organic material (mucks and peats). All of these variables will help in determining the quantity and kind of microorganisms associated with particular types of soils. (See Fig. I-3 for a profile of soil.)

Water

Water environments may vary similarly in their physical and chemical characteristics. Salt concentration is an obvious variable, as are temperature, pH, aeration, dissolved nutrients (organic and inorganic); still other variables are light, depth, suspended particles.

II

The Winogradsky Column

The Winogradsky column (see Fig. II-1) is a simple way to study a cross section of a natural environment in the laboratory. It represents a core-sampling of forest or field floor, freshwater or marine mudbank, pond, lake, stream, or bay bottom. The top of the column or core is exposed to air (oxygen); the further down in the column the less oxygen available until one reaches an intensely anaerobic zone. The top and sides of the column are exposed to light and permit the growth of a spectrum of organisms, from those that require light and oxygen to those that require light and are killed by oxygen. The column may be packed with soil, mud, and water from a variety of environments. It also lends itself to modification of the environment by enrichment at the time of packing the column or during incubation by the continuous or intermittent addition of enrichment materials from the top or bottom; the kind of radiant energy can also be varied.

A. Container

The vessel to contain the column can be made from transparent tubing which can be closed at one end, thin jars, cylinders, tubes, etc. The height or width of the container is not important. Clear plastic

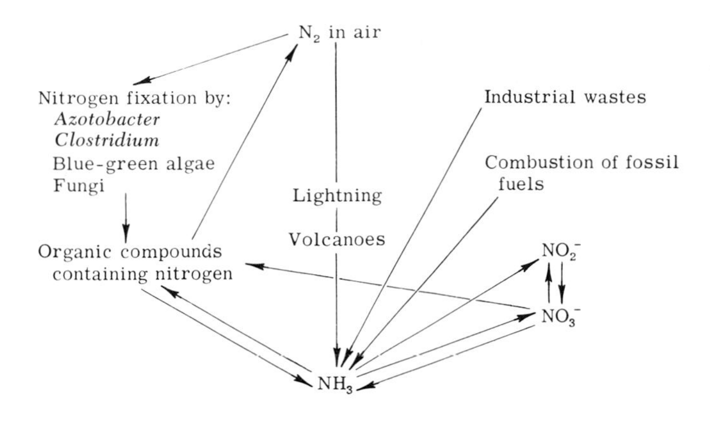

FIG. I-2. Nitrogen cycle.

sible for the unusually rich sulfur and iron deposits found in various parts of the world.

Microorganisms live and multiply in all the environments found on the earth: in soil; fresh, brackish, and marine water, at all depths; on ice and snow; on mountain tops; and in the air, where they may not multiply.

Soil

Soil is composed of mineral or organic matter, water, air, and organisms, the concentration of each varying with the type of soil and its source. Soils also vary in color, depth, and pH. Soils differ in texture; some soils contain large quantities of clay, others large quantities of sand or silt, while still other soils are rich in organic material (mucks and peats). All of these variables will help in determining the quantity and kind of microorganisms associated with particular types of soils. (See Fig. I-3 for a profile of soil.)

Water

Water environments may vary similarly in their physical and chemical characteristics. Salt concentration is an obvious variable, as are temperature, pH, aeration, dissolved nutrients (organic and inorganic); still other variables are light, depth, suspended particles.

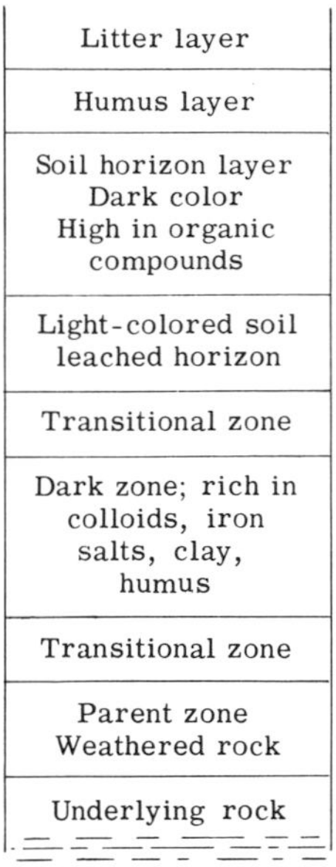

Litter layer
Humus layer
Soil horizon layer Dark color High in organic compounds
Light-colored soil leached horizon
Transitional zone
Dark zone; rich in colloids, iron salts, clay, humus
Transitional zone
Parent zone Weathered rock
Underlying rock

FIG. I-3. Profile of soil.

All of these factors will influence the quantity and kind of microorganisms found. (See Fig. I-4 for a brief description of aquatic environments.)

Microorganisms (bacteria, fungi, and algae) are found growing and multiplying on snow and ice, albeit slowly. Spores and resting stages and cells of all major groups of microorganisms have been found many miles in the air. It is likely that they are carried aloft by wind for there is no evidence that they grow or multiply there.

How does one study the variety of microorganisms in nature or isolate them from their natural environment for detailed study? Isolation *is necessary*; there are few observations that can be made on microorganisms in the field because of their small size. They have to be isolated in pure form on media specialized for their needs. This book concerns itself with the growth of a sample—soil, mud, shore, forest floor, fresh- or saltwater—in the laboratory or classroom and the proliferation by enrichment resulting in the isolation of the microorganisms desired.

General Classification of Lakes

Characteristics	Oligotrophic	Eutrophic
Basin	V-shaped	U-shaped or flat
Age	Young	Old
Depth	Deep	Shallow
Temperature	Cold	Warm
Bottom	Rocky	Silted
Margin (beach)	Little or none	Wide
O_2 content	High	Low
Suspended matter	Low	High
Dissolved organic material	Low	High
Dissolved inorganic material	Low	High
Microbial life	Low	High

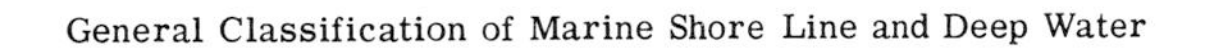
General Classification of Marine Shore Line and Deep Water

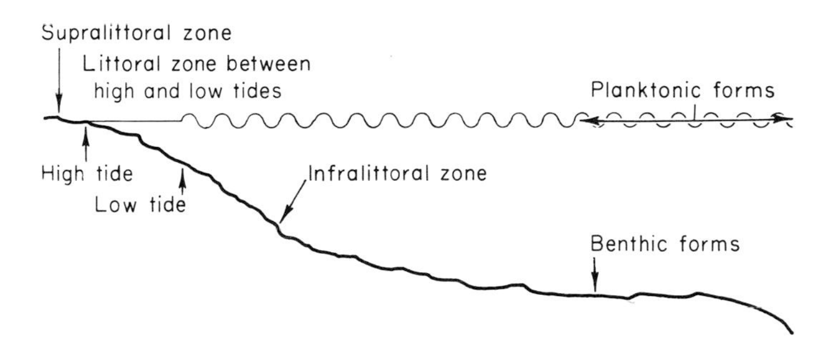

FIG. I-4. Description of aquatic environments.

II

The Winogradsky Column

The Winogradsky column (see Fig. II-1) is a simple way to study a cross section of a natural environment in the laboratory. It represents a core-sampling of forest or field floor, freshwater or marine mudbank, pond, lake, stream, or bay bottom. The top of the column or core is exposed to air (oxygen); the further down in the column the less oxygen available until one reaches an intensely anaerobic zone. The top and sides of the column are exposed to light and permit the growth of a spectrum of organisms, from those that require light and oxygen to those that require light and are killed by oxygen. The column may be packed with soil, mud, and water from a variety of environments. It also lends itself to modification of the environment by enrichment at the time of packing the column or during incubation by the continuous or intermittent addition of enrichment materials from the top or bottom; the kind of radiant energy can also be varied.

A. Container

The vessel to contain the column can be made from transparent tubing which can be closed at one end, thin jars, cylinders, tubes, etc. The height or width of the container is not important. Clear plastic

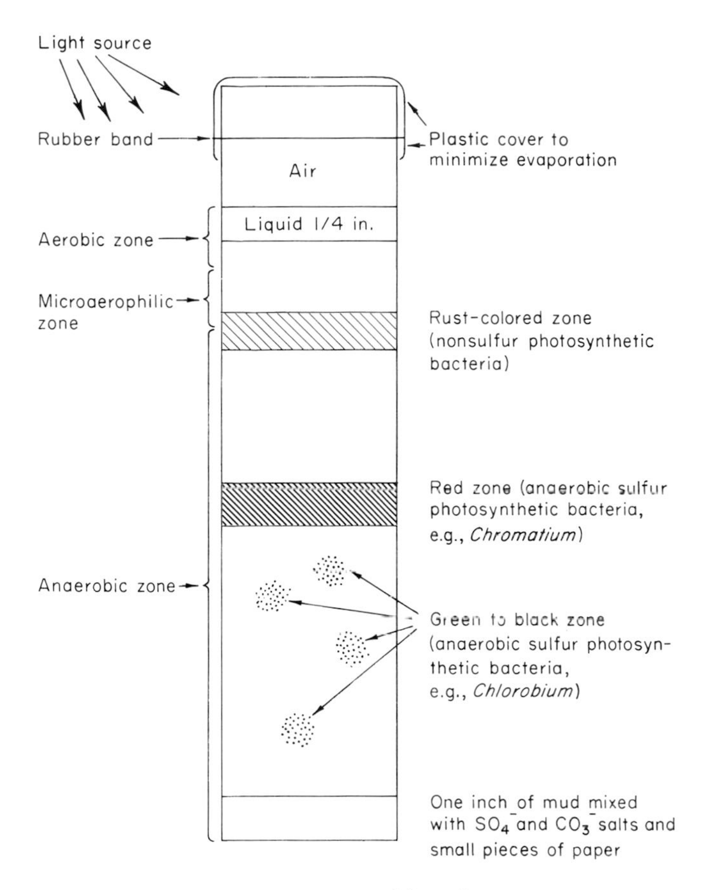

FIG. II-1. Winogradsky column.

as well as glass will do. The best dimensions, however, are a minimum of 1½-2 inches (3½-5 cm) wide and 6-8 inches (15-20 cm) high if you plan to remove soil with a spoon or spatula.

The column may be used with the continuous or intermittent trickle of marine or enriched or modified water (see Fig. II-2). For this purpose a column, i.e., chromatographic column with a large-pore sintered glass filter may be used or one with a rubber cork and a glass tube containing glass wool in the bottom of the column. These columns must not be packed tightly if fluids are to flow; the addition of sterile diatomaceous earth or fine sand (50%) by volume to the soil before packing the column improves drainage.

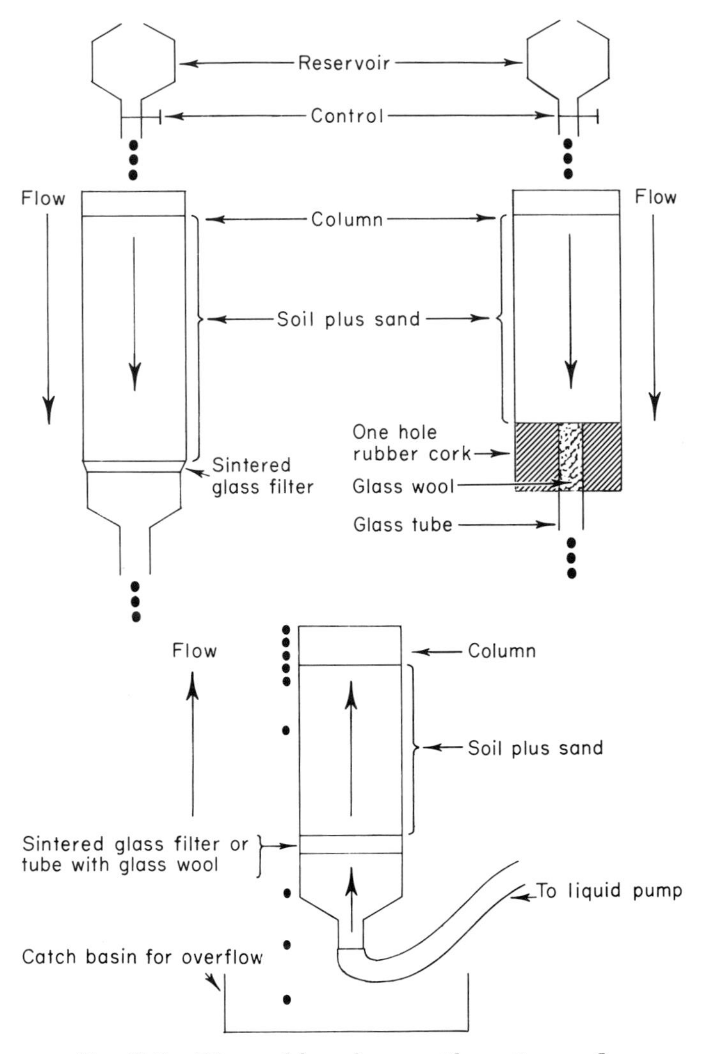

FIG. II-2. Winogradsky columns with continuous flow.

B. Choice of Soil

The best choice of soil for use in the column is the mud at the edge of a freshwater pond, lake, stream, river or a marine bay, canal, ocean, etc. (the "gooey-er" and smellier the better). Garden, field, or forest soil will also work and even the sandy shore of a beach has given adequate columns. If the water is too badly polluted the soil or mud is enriched for those microorganisms which can survive the pollutant. Even polluted waters and muds may yield interesting microorganisms.

C. Preparation of the Soil

If the soil is dry it is best to pulverize it to a powder and sift it through a strainer or collander to remove the larger pebbles, twigs, leaves, etc. Sifting the soil is recommended although not essential. Good column material may be prepared by removing the larger objects by hand.

Mud and wet soil is best prepared for the column by allowing it to stand in a tall container as a fluid with the consistency of heavy cream. This will allow the larger and heavier particles to settle. Light objects like twigs, leaves, etc. will have to be removed. The lighter supernatant fluids (liquid mud) is decanted into the container to be used for the column.

D. Enrichment of the Soil (or Mud)

It is customary to mix a small portion of the sifted soil or debris-free mud with a teaspoon of calcium carbonate and a teaspoon of calcium sulfate. These will provide an immediate source of sulfur and carbon dioxide as they dissolve and the products diffuse through the column. An extra long-term source of carbon and energy is usually provided by mixing finely shredded paper with the mud or soil. The paper is degraded to CO_2 and other small molecular products by microbial activity. If desired, any chemical may be added to the soil to create an environment which will select for those microorganisms which will resist the toxicity of the chemical, if it is poisonous, or which will even utilize the chemical for their metabolism.

E. Packing the Column

The mixture of mud (or soil) plus calcium sulfate and shredded paper is added to the bottom of the column container so that it occupies a height of about 1 inch (2.5 cm). If not already wet, it is then moistened with water (pond, salt, or distilled) to the consistency of light mud or heavy cream and mixed with a rod to insure proper consistency, uniformity, and wetness. Additional layers of wet soil (or mud) with a heavy cream consistency are added stepwise while continuing to mix the upper layers of added mud. The main purpose of this mixing is to eliminate air bubbles which might be trapped in the mud. The mud (or wet soil) is added to a mud column to a height of at least 4-6 inches (10-15 cm). Higher columns are permissible although a shorter column might not provide the depth for anaerobiosis which is necessary for growth of the anaerobic photosynthetic bacteria. The mud (or soil) in the column should be so well packed that there are no visible air bubbles; there should be about ¼ inch (0.6 cm) of clear water at the top of the column after the column has settled (12-24 hours). Excess water should be poured off. The column may be left uncovered but this permits evaporation: *the column must never be allowed to dry.* To minimize evaporation, the top of the column may be covered with a sheet of thin plastic (such as that used in the home for sandwiches) which is applied with a rubber band. The column should be examined periodically to see if the water layer is at ¼ inch or 0.6 cm; the height of the water layer may be varied.

F. Incubating the Column

The column is usually incubated at room temperature. Other temperatures may be chosen with the understanding that extreme temperatures select for microorganisms specialized to survive and multiply under these conditions. Cold-loving (psychrophilic) microorganisms will be selected for in the temperature range 0-20°C (32-68°F); thermophilic (heat-loving) microorganisms are the only ones that can multiply between 45-80°C (113-176°F). Most familiar organisms are mesophilic preferring a temperature between 20° and 45°C (68°-113°F).

Light for the column may be provided by any artificial illumination such as incandescent or fluorescent lamps. *Do not allow the lamps to heat the column.* Light from a window may also be used providing the column is not overheated by direct sunlight. Sunlit columns may take a week or two longer to develop their characteristic appearance.

G. What to Look For

After column has incubated for 3-6 weeks it will develop a characteristic appearance (see Fig. II-1), however, the appearance may be modified by the soil, growth conditions, or chemical additions.

H. Sampling the Column

1. The organisms in the column may be studied by direct observation by removing a sample of desired soil or water to a wet mount on a microscope slide (see techniques, Chapter V, p. 31) and examining the slide with the microscope. The best source of material for this type of study is the liquid layer at the top of the column or the soil immediately below it. These layers will contain the aerobic microorganisms: algae, protozoa, bacteria, fungi, and even metazoa.

2. To study the microorganisms throughout the column it is best to remove the liquid layer to another container. The remainder of the column is now carefully removed with a spoon or spatula in portions which are laid out on paper in the order in which they are removed (see Fig. II-3). Care is taken to note the portions containing the different anaerobic photosynthetic bacteria. Samples of the liquid at the top of the soil are removed with a bacteriological loop or needle (a bent dissecting needle may be used), and streaked onto or into the surface of the appropriate liquid or solid medium. The media and techniques for the isolation of different microorganisms are described in later sections. The procedure for streaking plates or stabbing butts are described on pp. 24, 25, and 35.

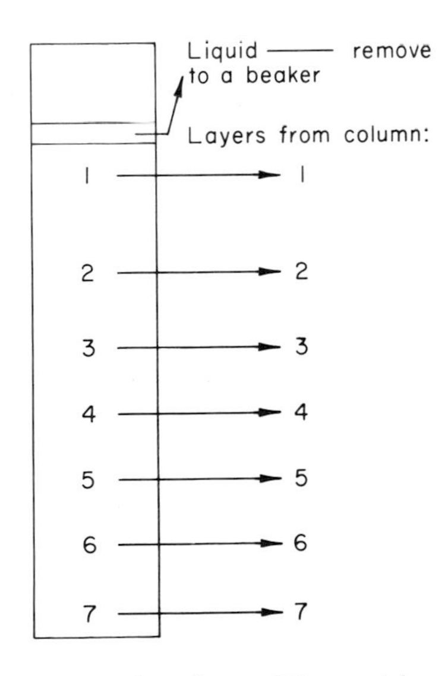

FIG. II-3. Sampling from a Winogradsky column.

III

Enrichment Technique

Enrichment ("elective") technique implies the creation of an environment which permits the desired organism to multiply at a greater rate than other organisms or which suppresses the growth of all but the desired organisms. The various procedures in general use are described below.

A. Successive Incubation

One may follow several procedures to select for microorganisms which prefer or demand a specific chemical or physical environment. In one method, one selects a specific environment (chemical and physical) and inoculates a sample of soil (mud or water) into the special environment. The sample is incubated for one or more weeks and an inoculum (loopful, pipetful, etc.) is then transferred to a new container of the same selected environment and incubated until the organism of choice appears (see Fig. III-1). This may require just a few or many transfers.

A second procedure is to gradually increase the concentration (if the selective factor is chemical) or to intensify the selective physical conditions starting with an inoculum of soil (mud or water) at a low concentration or noncritical physical level. After incubating for one

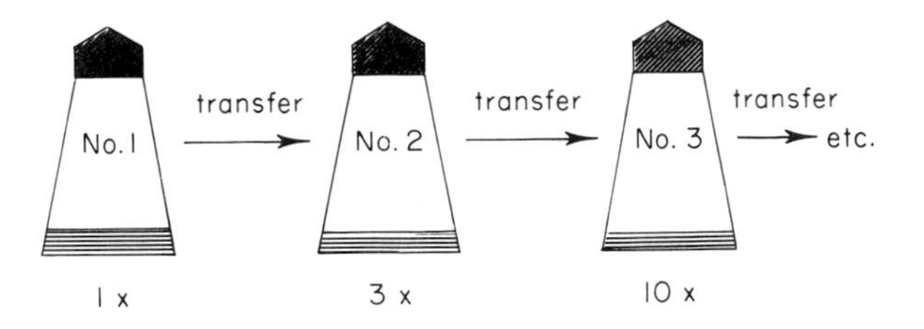

FIG. III-1. Enrichment or elective culture. Concentration or condition under investigation may be kept constant for several transfers or varied as shown from transfer to transfer.

or more weeks in the first enrichment culture, a sample (loopful, etc.) of this culture is transferred to a new container at a higher concentration or intensified physical condition. Again and again subcultures are transferred until the desired resistant or chemically specialized microorganisms are selected.

A third method for the enrichment of aquatic microorganisms is that of continuous culture. This method depends on competition for a limiting chemical compound. The constant flow of medium permits continuous multiplication and prevents the accumulation of cells and metabolic products which might nourish undesired organisms. Furthermore, the flow of nutrients aids the competitive process by diluting and removing the slow-growing species thus increasing the rate of selection for the fast-growing organism, that is, the one able to grow best in the selective medium. The details and equipment have been described by H. W. Jannasch [see *Arch. Mikrobiol.* **45**, 323–42 (1963); *ibid.* **59**, 165–73 (1967); *Limnol. Oceanogr.* **12**, 264–71 (1967)].

B. Choice of Media

1. *Minimal or Autotrophic Medium*

All microorganisms require a source of carbon. For most autotrophs this is best supplied as a carbonate salt or by bubbling carbon dioxide or air through the liquid medium. All microorganisms require nitrogen, magnesium, sulfur, phosphorus, sodium, potassium, and the trace elements iron, manganese, copper, cobalt, zinc, molybdenum, vanadium, boron, calcium, and probably also iodine, selenium, nickel, and chromium—and the list may be far from exhausted.

These trace elements may be supplied as individual soluble salts or they may be supplied as contaminants in the chemicals used in larger quantities. A selection of autotrophic media ranging from relatively simple to more complex will be found in the section on autotrophic media. Phosphate salts often precipitate and sodium glycerophosphate may be substituted for inorganic phosphates.

Many media precipitate if the salts are not added in a certain order or if chelating agents are lacking. Organic acids, bases, or amphoteric compounds may serve as chelating agents, i.e., amino acids, citrate, diamines; among the most effective chelating agents for most purposes, however, are those which are metabolically inert such as ethylenediaminetetraacetic acid (Versene) and nitrilotriacetic acid.

2. *Heterotrophic Media*

Many microorganisms require very few organic molecules to grow —perhaps only one or more vitamins, or a few amino acids, or only a carbohydrate as a carbon, and energy source. Many others demand elaborate organic recipes or complex media of natural origin. Those organisms requiring few or many organic compounds for growth are called *heterotrophs.* A good rule to remember is that the more organic material you add the more likely you are to enrich for the common microbes ("weeds") which quickly utilize media rich in readily metabolizable low-molecular nutrients while the more fastidious heterotrophs or the autotrophs are masked or swamped by these weeds. A number of mixtures of organic compounds are available to meet the needs of heterotrophs if added to the minimal medium in small judicious amounts. Judiciousness is an empirical idea. Some of these organic supplements with their major components are shown in Table III-1.

C. Growth Conditions

1. *pH*

The pH of choice is often, but not always, that of the environment in which the microorganism is normally found. If this is not known, one must surmise the pH of media or environment likely to support the desired microbe. In general most microorganisms, except for those with special needs or unusual abilities, do best in the pH range 6–8. Buffers may be added to maintain pH if there are changes during growth.

TABLE III-1
ORGANIC SUPPLEMENTS

Supplement	Description
Soybean lecithin	Rich in lipids such as fatty acids, sterols, etc.
TEM (diacetyl tartaric ester of tallow glycerides)	Rich in lipids such as fatty acids, sterols, etc.; water soluble
Tween 40 (polyoxyethylene sorbitan ester of palmitic acid)	Rich in lipids such as fatty acids, sterols, etc.; water soluble
Tween 60 (polyoxyethylene sorbitan ester of stearic acid)	Rich in lipids such as fatty acids, sterols, etc.; water soluble
Tween 80 (polyoxyethylene sorbitan ester of oleic acid)	Rich in lipids such as fatty acids, sterols, etc.; water soluble
Gelatin hydrolyzate	Rich in amino acids except L-methionine and L-tryptophan
Trypticase and similar enzymatic casein digests	Rich in amino acids and water-soluble vitamins
Hycase (acid digest of casein)	Rich in amino acids and water-soluble vitamins
Proteose peptone	Rich in amino acids and water-soluble vitamins
Other peptones	Rich in amino acids and water-soluble vitamins
"Liver L"	Rich in water soluble vitamins, pyrimidines, purine, amino acids
Soil extract	Rich in minerals, water-soluble vitamins, organic compounds of great variety

2. *Temperature*

The temperature of choice is again, but not always, that of the microbe's natural environment when known. If not known, it is reasonable to use 20°–30°C for most microorganisms. Psychrophilic forms prefer temperatures below 20°C while those requiring high temperatures will do best between 45°–55°C (some blue-green algae go as high as 74°C; some bacteria may go higher).

3. *Anaerobiosis*

Microorganisms vary in their need for atmospheric oxygen. Some will only grow at surfaces of liquids or solid media where the oxygen supply is plentiful, others grow only with minimal oxygen—fairly deep in liquid or solid media; while still others cannot tolerate even traces of oxygen and will grow only at the bottom of solid media or under growth conditions rigorously excluding oxygen. Many microorganisms are facultative and can grow in the presence or absence of oxygen; these are the ones most likely to multiply. Special methods are available to exclude oxygen for those microbes not tolerating it; these are described in a special section (p. 34).

D. Soil Extract

This is the "elixir of life" for the growth of many microorganisms when they are first isolated, since it contains all the inorganic and organic compounds located in the soil or water in which many microbes are found. Soils vary as do their microbial flora and fauna and an extract of one soil may not always produce desired results. If this happens extract other soils. Also if the soil extract—or for that matter other materials—is prepared by heating, this will at once destroy heat-labile nutrients and tend to eliminate microorganisms that require them. Soil extract should be used at low concentrations; higher concentrations are often toxic.

E. Inorganic Enrichments

Many microorganisms, particularly bacteria, require only light or a reduced inorganic molecule as an energy source; carbon can be supplied by CO_2 or carbonates (CO_2 is lost readily from acid media). Nitrogen can be supplied as ammonium or nitrate salts, sulfur as sulfates, and phosphorus can be supplied as phosphates which are often insoluble but may be replaced by glycerophosphate salts. The remaining trace elements (iron, zinc, magnesium, molybdenum, copper, cobalt, vanadium, boron) can be supplied as special trace element supplements or as contaminants in the bulk salts. Examples of the trace element supplements are given in a later section. The quantity of any one of the following salts: phosphate, nitrate, and sulfate used in inorganic media seems to select for some types of photosynthetic microorganisms. Examples of enriched seawater media are given in Table III-2. "Aged" seawater is ordinary seawater which has been stored in the dark for months or years—presumably permitting complete mineralization to occur.

TABLE III-2
SEAWATER ENRICHMENTS[a]

Nutrient	Schreiber (1927)	Foyn's Erd-Schreiber (1934a)	Barker (1935)	Sweeney (1951)	Goldberg *et al.* (1951)	Spencer (1954)	Sweeney (1954)	Haxo and Sweeney (1955)	Wilson and Collier (1955)
$NaNO_3$	10 mg	10 mg	–	–	–	Varied	–	–	–
KNO_3	–	–	10 mg	20.2 mg	20.2 mg	–	20.2 mg	20.2 mg	–
NH_4Cl	–	–	–	–	–	–	–	–	0.1 mg
$Na_2HPO_4 \cdot 12\ H_2O$	2 mg	2 mg	–	–	–	Varied	–	–	–
K_2HPO_4	–	–	1 mg	3.5 mg	Varied	–	3.5 mg	3.5 mg	–
KH_2PO_4	–	–	–	–	–	–	–	–	0.05 mg
$FeCl_3$	–	–	0.01 mg	0.1 mg	–	–	0.097 mg	0.097 mg	–
Fe Citrate	–	–	–	–	0.24 mg	0.01 mg	–	–	–
$MnCl_2$	–	–	–	0.0075 mg	0.013 mg	0.0045 mg	0.0075 mg	0.0075 mg	–
Na_2HAsO_4	–	–	–	–	0.018 mg	–	–	–	–
$CoCl_2$	–	–	–	–	0.013 mg	–	–	–	–
$MgCl_2 \cdot 6\ H_2O$	–	–	–	–	–	–	–	–	0.02 mg
$NaHCO_3$	–	–	–	–	–	–	–	–	0.1 mg
$Na_2S \cdot 9\ H_2O$	–	–	–	–	–	–	–	–	0.1 mg
EDTA	–	–	–	–	–	–	1 mg	1 mg	–
B_{12}	–	–	–	–	–	–	1 μg	–	0.1 μg
Thiamine·HCl	–	–	–	–	–	–	–	–	1.0 mg
Biotin	–	–	–	–	–	–	–	–	0.05 μg
Soil extract	–	5 ml	2 ml	4 ml	–	–	–	2 ml	2 ml
Seawater	100 ml	100 ml	–	–	–	–	–	–	–
Aged seawater	–	–	100 ml	100 ml	100 ml	75 mg	75 ml	75 ml	95 ml
Distilled water	–	–	–	–	–	25 mg	25 ml	25 ml	5 ml

[a]From L. Provasoli, J. J. A. McLaughlin and M. Droop, *Arch. Mikrobiol.* **25**, 392 (1957).

F. Artificial Seawater

Synthetic salt mixtures have been compounded to simulate seawater. Some have proven useful. None are certain to give good results with all marine microorganisms. It is best to use natural seawater when first isolating new organisms. After isolation the ability of artificial seawaters to support the growth and multiplication of these organisms may be assayed. Examples of these media are given in Table III-3, and III-4. A synthetic sea salt mixture "Instant Ocean" may be purchased from Aquarium Systems, Inc. 1450 East 289 Street, Wickliffe, Ohio 44092.

G. Organic Enrichments

Any mineral or organic freshwater or marine medium may be enriched with a variety of organic compounds. It is best, however, if one is seeking to isolate an organism capable of metabolizing a particular organic compound to have no other major organic molecule present. If this is not possible in the presence of, for example, soil extract, vitamin mix, it is nevertheless necessary to minimize the amount of organic material present so as to force the selection of those microorganisms which can utilize the desired molecule for carbon and energy. Bacteria and fungi lend themselves to this type of enrichment; a large variety have been isolated capable of metabolizing a host of unusual organic compounds. Among such specialists are organisms using naphthalene and other petroleum constituents, and such refractory, though common natural polymers, as lignin and chitin.

H. Selective Inhibitors

1. Antimetabolites

A variety of metabolic analogs (compounds resembling normal metabolites) have been synthesized which interfere with the metabolism of the normal metabolites. Some occur in nature. These compounds may be used to select for microorganisms which (*a*) lack the metabolic pathway blocked by the antimetabolite or use an alternative pathway or (*b*) have unusual metabolic properties re that specific antimetabolite, i.e., resistance, lack of permeability, ability to use compound. Among the compounds available are analogs of almost all

TABLE III-3
SELECTED SYNTHETIC MARINE MEDIA
(HASKINS LABORATORIES)[a]

Nutrient	ASP	RC	DC	ASP 2	ASP 6
NaCl	2.4 g	2.1 g	1.8 g	1.8 g	2.4 g
$MgSO_4 \cdot 7\ H_2O$	0.6 g	–	0.5 g	0.5 g	0.8 g
$MgCl_2 \cdot 6\ H_2O$	0.45 g	0.5 g	–	–	–
KCl	0.06 g	0.06 g	0.06 g	0.06 g	0.07 g
Ca (as Cl^-)	40 mg	7 mg	10 mg	10 mg	15 mg
$Na_2SO_4 \cdot 10\ H_2O$	–	0.3 g	–	–	–
$NaNO_3$	–	–	50 mg	5 mg	30 mg
KNO_3	10 mg	10 mg	–	–	–
K_2HPO_4	2 mg	1 mg	–	0.5 mg	–
K_2 glycero-phosphate	–	–	40 mg	–	10 mg
$Na_2SiO_3 \cdot 9\ H_2O$	2.5 mg	–	20 mg	15 mg	7 mg
"Tris"	–	0.5 g	0.5 g	0.1 g	0.1 g
Thiamine	–	0.1 mg	–	–	–
Biotin	–	0.05 μg	–	–	–
B_{12}	0.02 μg	0.1 μg	0.3 μg	0.2 μg	0.05 μg
Vitamin mix 8[b]	0.05 ml	0.02 ml	0.0 ml	–	0.1 ml
Vitamin mix S3[c]	–	–	–	1.0 ml	–
Na taurocholate	–	0.3 mg	–	–	–
Na lactate	–	–	0.05 g	–	–
Na acetate·3 H_2O	–	0.02 g	0.05 g	–	–
Na H glutamate	–	0.05 g	0.05 g	–	–
Glycine	–	–	0.05 g	–	–
Sucrose	–	0.07 g	0.05 g	–	–
Na_2EDTA	1.0 mg	1.0 mg	3.0 mg	3.0 mg	–
Na_3 Versenol	–	–	–	–	3.0 mg
Fe (as Cl—)	0.01 mg	0.01 mg	0.08 mg	0.08 mg	0.2 mg
Zn (as Cl^-)	5.0 μg	5.0 μg	15.0 μg	15.0 μg	0.05 mg
Mn (as Cl^-)	0.04 mg	0.04 mg	0.12 mg	0.12 mg	0.1 mg
Co (as Cl^-)	0.1 μg	0.1 μg	0.3 μg	0.3 μg	1.0 μg
Cu (as Cl^-)	0.04 μg	0.04 μg	0.12 μg	0.12 μg	2.0 μg
B (as H_3BO_3)	0.2 mg	0.2 mg	0.6 mg	0.6 mg	0.2 mg
"1S" metals[d]	–	1.0 ml	–	–	–
Mo (as Na salt)	–	–	–	–	0.05 mg
H_2O	100 ml	100 ml	100 ml	100 ml	100 ml
pH	7.6	7.2-7.4	7.6-8.0	7.6-7.8	7.4-7.6

[a]From L. Provasoli, J. J. A. McLaughlin and M. R. Droop, *Arch. Mikrobiol.* **25**, 392 (1957).

[b]1 ml of Vitamin mix 8 contains: thiamine · HCl, 0.2 mg; nicotinic acid, 0.1 mg; putrescine·2 HCl, 0.04 mg; Ca pantothenate, 0.1 mg; riboflavin, 5.0 μg; pyridoxine·2 HCl, 0.04 mg; pyridoxamine·2 HCl, 0.02 mg; *p*-aminobenzoic acid 0.01 mg; biotin, 0.5 μg; choline·H_2 citrate, 0.5 mg; inositol, 1.0 mg; thymine, 0.8 mg; orotic acid, 0.26 mg; B_{12}, 0.05 μg; folinic acid, 0.2 μg; folic acid, 2.5 μg.

[c]1 ml of Vitamin mix S3 contains: thiamine·HCl, 0.05 mg; nicotinic acid, 0.01 mg; Ca pantothenate, 0.01 mg; *p*-aminobenzoic acid, 1.0 μg; biotin, 0.1 μg; inositol, 0.5 mg; folic acid, 0.2 μg; thymine, 0.3 mg.

[d]1 ml of "1S" metals contains: Sr, 1.3 mg; Al, 0.05 mg; Rb, 0.02 mg; Li, 0.01 mg; I, 0.005 mg; Br, 6.5 mg.

TABLE III-4
ARTIFICIAL SEAWATER[a]

Nutrient	Quantity
NaCl	2.348 g
Na_2SO_4	0.392 g
$NaHCO_3$	0.019 g
KC1	0.066 g
KBr	0.0096 g
H_3BO_3	0.0026 g
$MgCl_2 \cdot 6\ H_2O$	1.061 g
$SrCl_2 \cdot 6\ H_2O$	0.004 g
$CaCl_2 \cdot 2\ H_2O$	0.1469 g
Water to	100 ml
pH	7.2

[a]From P. Burkholder. *In* "Symposium on Marine Microbiology" (C. H. Oppenheimer, ed.), pp. 133–150. Thomas, Springfield, Illinois, 1963.

the low-molecular weight compounds involved in metabolism: vitamins, purines, pyrimidines, amino acids, etc. Examples of some metabolic analogs which are available commercially are given in Table III-5.

2. *Selection of Microorganisms which Secrete and Perhaps Excrete Metabolites*

A method was developed by E. A. Adelberg, [*J. Bacteriol.* **76**, 326 (1958)] for isolating mutants whose resistance to antimetabolites depends on their synthesis of unusually high concentrations of the target metabolite, thereby overcoming inhibition by the antimetabolite. Another method which may be used to isolate metabolite-secreting microorganisms is bioautography. Several of these procedures are detailed in Chapter VII, p. 167.

3. *Antibiotics*

These products of microbial metabolism inhibit the growth or metabolism of other microorganisms. The classical example is penicillin which inhibits microorganisms by interfering with the synthesis of cell walls containing muramic acid. Many other specific antibiotics are available; they may be used to inhibit the multiplication of

TABLE III-5
SOME ANALOGS OF METABOLITES

Metabolite	Analog (antimetabolite)
Vitamins	
Nicotinic acid	3-Acetylpyridine
Biotin	Avidin
	Desthiobiotin
Ascorbic acid	Dehydroascorbic acid
Vitamin K_1	Dicumarol
Nicotinic acid	Nicotinuric acid
Thiamine	Neopyrithiamine
	Oxythiamine
Pantothenic acid	Pantoyl taurine
Nicotinic acid	α-Picolinic acid
	Pyridine-3-sulfonic acid
p-Aminobenzoic acid	Sulfanilamide
Amino acids	
L-Cysteine	Allylglycine
Leucine	4-Azaleucine
Phenylalanine	DL-β-Phenyllactic acid
Arginine	Canavanine
Methionine	Ethionine
Tryptophan	Indole
Phenylalanine	DL-*p*-Fluorophenylalanine
Purines	
Adenine	Benzimidazole
	Azaadenine
Guanine	Caffeine
	Azaguanine
	2,6-Diaminopurine
Pyrimidines	
Thymine	5-Fluorouracil
	5-Bromouracil
	6-Azathymine
Uracil	6-Azauracil
	Diazouracil
	Thiouracil
	Barbituric acid

undesirable microorganisms in enrichments, or to wash isolated organisms free from contaminating organisms, or permit them to outgrow their contaminants. A list of antibiotics and microorganisms most sensitive to them is included in Table III-6. Some antibiotics have relatively specific sites of inhibition which may be exploited. Some of these are listed in Table III-7.

TABLE III-6
ANTIBIOTICS

Antifungal	Antibacterial (probably anti-blue-green algal)
Amphotericin B	Bacitracin
Candicidin	Chloramphenicol
Cycloheximide (actidione)	D-Cycloserine
Filipin	Erythromycin
Griseofulvin	Neomycin
Mycostatin	Novobiocin (cathomycin, albamycin)
Nystatin	Penicillin
Pimaricin	Polymyxin B and E
Rimocidin	Streptomycin
Antiprotozoal	Tetracycline
Cycloheximide (actidione)	Tyrothricin
Antialgal	
Cycloheximide (actidione)	

TABLE III-7
ANTIBIOTICS

Inhibition of	
DNA synthesis	Bacterial cell wall synthesis
Mitomycin	Penicillin
Phleomycin	D-Cycloserine (oxamycin)
Porfiromycin	Vancomycin
Protein synthesis	RNA synthesis
Chloramphenicol	Actinomycin
Cycloheximide (actidione)	Mithramycin
Erythromycin	Olivomycin
Puromycin	Membrane function
Streptomycin	Amphotericin C
Tetracyclines	Filipin
	Mycostatin (nystatin)

IV

Isolation Procedures

There are many ways to isolate microorganisms. Different methods are needed for different organisms. Some microorganisms are easy to isolate in "axenic" culture, that is, free of other organisms. Other microorganisms may require the presence of one or more microorganisms which provide essential nutrients and these may only be removed after the nutrients are supplied. Some of the methods presented here are used in the author's laboratory as well as in other laboratories. The reader's ingenuity may improve them. Microorganisms in soil or mud are usually suspended in a sterile liquid medium before attempting isolation.

A. Dilution Series

Liquid suspensions of the desired microorganism may be diluted in sterile liquid medium until their numbers, as well as those of their contaminants, are reduced. If samples of the diluted suspensions are then streaked on special agar media or inoculated into liquid media the desired microorganism may multiply free of its contaminants (axenic culture). In some instances the original or diluted suspensions may be examined microscopically and the desired microorganism picked out with a micropipette and placed in a sterile medium (see "washing methods," p. 28).

B. Surface Plate Method

Liquid or solid samples containing desired microorganisms may be picked up with a sterile inoculating needle or loop and streaked across the surface of a selected solid medium in any one of the patterns seen in Fig. IV-1. Do not dig into the medium. If this streaking is properly done, the desired microorganisms may multiply free of contaminants. A drop of liquid containing desired microorganisms may be placed on a solid medium and then wiped across the surface with a sterile L-shaped glass rod. This method may be used for a crude dilution series by streaking surface in limited areas or quadrants, then sterilizing between transfers of material between the already streaked area (or quadrant) and the next area (or quadrant) to be streaked (see Fig. IV-1).

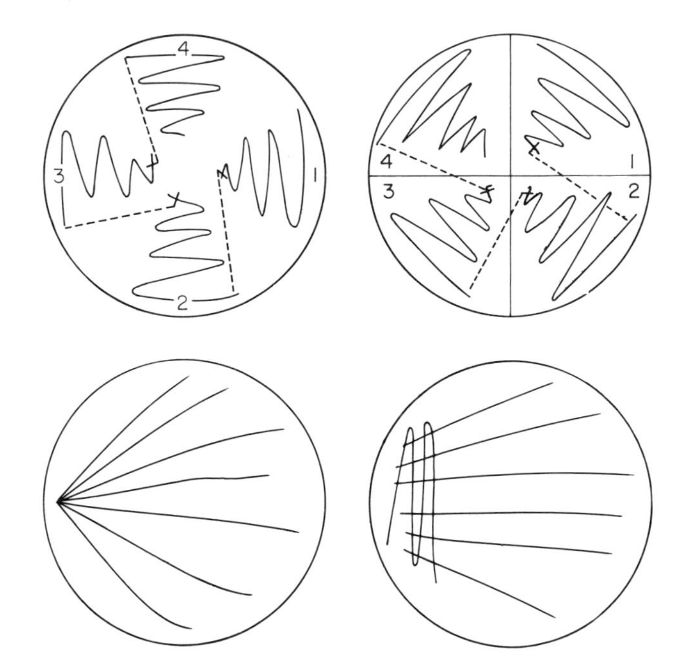

FIG. IV-1. Surface agar streak method for the isolation of pure cultures. Use needle or L-shaped glass rod. *Do not dig in the agar.* The solid lines represent the areas of contact between the needle and agar. Flame needle and cool before changing direction of streak shown by dotted lines. *Remember:* cool needle before touching it to last streak and starting new streak.

C. Agar-Shake Method

The liquid or solid sample is mixed with melted nutrient agar which has been allowed to cool to a temperature just short of hardening. Mixing may be done by shaking melted agar or by rolling the tube between the palms. Inoculated melted agar may be poured into sterile Petri plates or left in the tube to harden. Microorganisms may multiply in a well-isolated area of plate or tube so that they may be easily separated from their contaminants. Tubes containing isolated colonies may be cut with glass cutter sterilely to remove colonies or they may be picked out with a sterile, stiff wire loop or spatula.

This method also lends itself to a crude type of dilution series by using a sterile rod, needle, or loop to transfer material from one agar-shake tube to another tube before allowing agar to harden.

Agar may be prevented from hardening if kept in a water bath at 45°-50°C before and during use. However, these relatively high temperatures may kill many microorganisms.

D. Streak Feeding

Some protozoa (amebae, small flagellates) slime bacteria, and slime molds may be isolated free from contaminating microorganisms by streaking their food organisms, usually a bacterium (preferably dead), over the surface of an agar plate. A sample of liquid, soil, etc. containing these organisms is placed in the center of the plate. The desired motile organisms will find its food organism, and following the streak as it feeds may migrate away from some, if not all, contaminants. The progress of this baiting technique may be followed under a binocular dissecting microscope if the organisms are large enough. The atmosphere over the agar surface must be kept moist. A humid chamber for the incubation of the plates will do or the plates may be sealed with Parafilm or covered with a plastic sandwich bag and sealed.

E. Overlay Method

In this method microorganisms are streaked out over the surface of a solid nutrient medium and then permitted to multiply. When they have multiplied to the point where they are visible to the naked eye, a thin layer of nutrient or nonnutrient agar is poured over the surface of the agar plate and the colonies of microorganisms. Incubation is then continued. Some microorganisms may grow or move through the thin overlay of agar to form new surface cultures free of contaminants which they have either moved away from or rubbed off in their migration through the agar. This type of growth or movement may be encouraged by presenting light at the new surface for photosynthetic organisms, air for aerobic ones, or supplying discs soaked in required nutrients at the new surface, for motile heterotrophic forms. The use of chemotactic agents for this purpose is almost completely unexplored.

F. Filter-Disc Method

1. *Motile Microorganisms*

Liquid samples containing the desired microorganism may be passed through any one of several commercial filter discs with a pore size appropriate to the size of the organism. After filtration the discs may be placed on the surface of an appropriate sterile nutrient agar with the upper surface of the filter disc facing upward for several hours or for enough time to allow motile microorganisms to move through the pores of filter disc onto the nutrient agar surface. The disc is then removed. Microorganisms which have passed through the filter disc are now cultured free of most, if not all, of their contaminants.

2. *Nonmotile Microorganism*

Liquid samples containing the desired microorganism in small numbers are filtered through a filter disc of a pore size which will prevent their passage. The disc is placed on the surface of nutrient agar medium, upper face down, for about ½ hour. The disc is removed carefully to avoid smearing the agar surface and the agar plate is now incubated to allow growth of the isolated cultures of the desired microorganism. This method may also be used by smearing the disc over the surface of agar.

G. Phototaxis

Microorganisms which respond positively or negatively to light may be concentrated and sometimes made almost contaminant-free by exploiting their movement toward or away from a light source. In a medium made viscous with agar, motile microorganisms may be induced to swim away from their contaminants, rubbing them off as they move through the agar. This idea is also applicable to forms which grow through agar toward light.

H. Centrifugation

Many microorganisms and their cysts, spores, etc., may be separated from contaminating organisms and concentrated for further treatment by centrifugation. Depending on their density some are packed down; others float. This density difference may be more subtly exploited after initial concentration by centrifuging the desired cell material in a density gradient in which the desired microorganisms will seek a level corresponding to their own density while their contaminants, which may be only slightly different, are washed off, and find their own density levels. The density gradient may be discontinuous or continuous and is usually made with biologically inert molecules, i.e., sucrose, pentaerythritol, mannitol, Ficol. A continuous gradient is one in which the density varies by very small differences from a low concentration at top to a high concentration at the bottom of the centrifuge tube. This requires a special gradient-making device. A discontinuous gradient is one in which a series of solutions of varying concentration (density) are overlaid one on another: densest at the bottom, next dense on top, and so on. This can be easily made by carefully pipetting a few milliliters of the most concentrated solution into a centrifuge tube, then carefully, one after another, equal volumes of material of decreasing concentration. It is sometimes necessary to let the second and subsequent layers slowly run down the side of the tube to avoid mixing with the layers already present.

The centrifuge method may also be used in conjunction with agents which kill or remove the contaminants such as antibiotics, disinfectants, antiseptics, mild detergents, etc., or agar or gelatin which rubs off contaminants as the desired microorganisms are forced through the viscous material to be collected at the bottom of the centrifuge tube.

I. Ultraviolet Light

Short-wave ultraviolet light (250–300 mμ) is both mutagenic and lethal, but in a few instances, it has been found useful in separating microorganisms, especially blue-green algae, from their more vulnerable contaminants. Trials of time of exposure and distance from the ultraviolet source are necessary to define the conditions which kill many, but not all, of the desired microorganisms while killing all their contaminants.

J. Sonication and Cavitation

Several instruments are available which will induce supersonic vibrations in liquids (sonicators, ultrasonic instrument cleaners) or shearing forces in liquids (blendors, high-speed stirrers, etc.). Both kinds of instruments may be used to shake or wash contaminants off or away from desired microorganisms providing they do not tear apart the desired cells. These methods are useful for relatively sturdy cells and in conjunction with one or more of the other methods. Care must be taken to cool the sample or use it for only brief intervals since both methods generate much internal heat which may kill the organisms.

K. Washing Procedures

These are among the most tedious but most successful methods developed for preparing axenic cultures of various relatively large microorganisms. They entail successive passages of the desired microorganisms through sterile solutions which may contain antibiotics germicides, detergents, etc., against other microorganisms. The methods depend on dilution and the differential destruction of undesirable organisms. This method may also make use of the ability of motile organisms to swim away from some or all of their contaminants. The procedure may involve serial dilution in sterile solution or carefully picking the organism with a sterile micropipette or loop and placing them in a drop of sterile or inhibitor-containing liquid in a well of a multiwell glass dish, then carefully transferring them from well-to-well through many washings until they are free of contaminants—an empirical matter at best—and allowing them to grow in the nutrient medium chosen. It is not uncommon to find that some microorganisms will no longer grow or multiply in the absence of their contaminants which supplied essential nutrients. These organisms can only be maintained with their contaminants until their essential nutrients have been identified and then supplied.

The washing solutions to remove various contaminants are as

varied as the contaminants. They fall into categories which may be combined: among them are antibiotics, germicides (antiseptics and disinfectants), detergents, and enzymes. The concentrations of these agents to be used varies with the sensitivity of the desired organism and its contaminants and are best determined empirically providing there is understanding of the general nature of the inhibitory or destructive agent (see Tables III-5, 6, and 7; Table IV-1).

L. Geotaxis

Advantage may be taken of the fact that some microorganisms migrate to the bottom of long tube (positive geotaxis); others move upward (negative geotaxis). These tropisms may be used to concentrate and even axenize a desired species by having the microorganisms move through dilute agar (0.2–0.5%) with or without added antibiotic mixtures.

M. Migration through V-Tubes

Motile organisms placed at the top of one arm of a V-shaped tube containing fluid may be induced or will randomly swim into other arm and perhaps away from contaminating organisms. This method may also be used with dilute agar and antibiotic mixtures.

TABLE IV-1
GERMICIDAL AGENTS, DETERGENTS USEFUL FOR ISOLATING MICROORGANISMS

Disinfectants	Detergents and soaps
Acriflavines (2–3%)	Duponol
Chlorox (dilute)	Roccal
Gentian violet (2–3%)	Salts of fatty acids
Hexachlorophene (2–3%)	Sodium lauryl sulfate
Hypochlorite salts (dilute)	Zephiran
Lysol (5%)	Enzymes
Sulfanilamide	Cellulases
Tincture of iodine (2%)	Chitinases
Tincture of Merthiolate or Mercurochrome	Muramidase (lysozyme)-antibacterial
Detergents and soaps	Pronase
Bile salts	Snail "cytase"
	Trypsin, chymotrypsin

V

Some Microbiological Techniques

This section deals with some of the techniques that must be mastered in order to work with microorganisms. Only the more important techniques or representative ones are presented here. The reader is advised to read the references recommended for specific groups of microorganisms for detailed procedures or references to other methods.

A. Hanging Drop Method for the Examination of Microorganisms

1. Clean a depression slide and several cover slips. (Wash in soap and water; rinse well; dip into 95% ethanol; dry.)
2. Place a thin ring of petroleum jelly (Vaseline) around the depression.
3. Mark a thin ring around the rim of the coverslip with a crayon pencil. Don't press down on the cover slip as it cracks easily. Make sure the working surface is smooth.
4. Place a loopful of a culture of the microorganism to be studied in the center of the ring on the cover slip.
5. Invert depression slide over the cover slip, with the well or depression over the drop. Press down gently. Quickly invert so that the cover slip is on top.

6. Examine under microscope (see Fig. V-1).
7. Focus on the crayon ring to find the correct optical level and then move to the drop.
8. Make sure there is not too much light.
9. Microorganisms will appear to move in a definite direction. The jiggling movement common to all small particles seen in a liquid is *Brownian motion.*

B. Wet Mount

A drop of liquid to be examined for microorganisms is placed on a clean slide. Cover the drop with a clean cover slip. Examine under the microscope. Many microorganisms are more easily seen in dim rather than bright light. *Wet mounts dry rapidly.*

C. Slowing Down Motile Microorganisms

The following procedures may be used to slow down rapidly moving microorganisms in wet mounts:

1. One or more drops of methyl cellulose (10% in water).
2. Tangle of fine cotton threads or fibers from lens paper placed in drop.

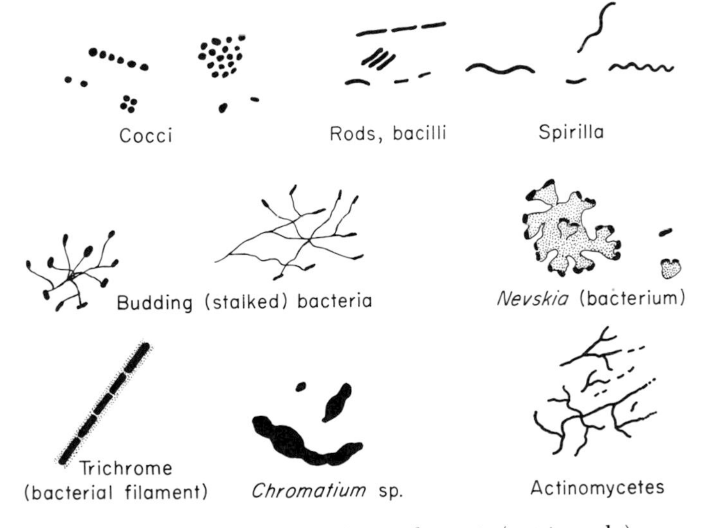

FIG. V-1. Appearance of some bacteria (not in scale).

3. For some protozoa, movement may be stopped by adding a drop of novocaine (1%) or cocaine hydrochloride (1%) to the drop containing the protozoa.

D. Aseptic Technique

Glassware, media, transfer materials, etc., all must be sterilized if microorganisms are to be isolated and maintained in single-species (axenic, "pure") culture. The aseptic techniques of microbiology must be used. These include:

(1). Minimizing or, preferably, eliminating dust or other sources of contamination. Germicidal ultraviolet lamps are useful over inoculating bench area. The inoculation area must be clear. Wash with germicide before use. Avoid stirring up dust in the inoculating room.

(2). Transfer microorganisms with sterile equipment to sterile glassware and equipment unless it is obvious that nothing is to be gained. See next section and microbiology texts and manuals for details on sterilization procedures.

E. Sterilizing Procedures

The sterilization procedure to be used depends on equipment available and the materials to be sterilized. Liquid media, plastic articles, etc., cannot be sterilized by dry heat; others like serum, embryo extract, and most antibiotics are unstable at even low temperatures and must be filtered. Glassware may be sterilized by any one of several methods.

1. Boiling and Steaming

Stable materials of glass or metal which can be completely immersed in boiling water may be sterilized by ½ to 1 hour in boiling water. If total immersion is not possible, i.e., media in glass containers, the material may be placed over a boiling water bath and steamed for 1 hour on each of 3 successive days.

2. Autoclaving (Autoclave or Pressure Cooker)

This is the method of choice for moist materials which are heat stable. Depending on size and volume as little as 15 minutes at 121°C (15 psi of steam pressure) suffices to sterilize a few small objects. Densely packed objects like tubes in a basket or large volumes of liquid medium may require 1 or more hours under the same conditions (15 psi and 121°C). All screw caps, closures, etc., must be sealed loosely if steam is to enter.

3. Filtration

Several devices are available which can be sterilized by autoclave or boiling water and can then be used to filter-sterilize liquids which cannot be sterilized by other means. The older filtration methods include Seitz asbestos filters, porcelain filters, sintered-glass filters; the newer methods include membrane filters. Appropriate holders and membranes are fabricated by: Gellman Instruments, 600 S. Wagner Rd., Ann Arbor, Michigan 48106; Millipore Corp., Bedford, Massachusetts 01730; Carl Schleicher & Schull, 543 Washington, Keene, New Hampshire 03431.

4. Gases

These are available from most laboratory supply houses. Ethylene oxide or β-propiolactone may be introduced into a closed container to sterilize materials which are not affected by them. The gas must be removed before the material may be used. The time of exposure varies with the materials to be sterilized and the gas used.

5. Dry Heat

Materials not changed or destroyed by high temperature, i.e., glass, may be sterilized in an oven for 2 hours at 165°C.

6. Ultraviolet Light

Materials not affected by ultraviolet light such as glass may be surface-sterilized by ultraviolet light as long as one takes into account that it does not penetrate most materials.

7. Chemicals

Various chemicals may be used to kill most microorganisms on glassware and other equipment. These include ethanol (70%), 5% sodium or calcium hypochlorite (Chlorox), iodophors (Wescodyne, Iodide, etc.), $HgCl_2$ (1:1000–1:5000), phenol (2–5%). Ethanol may be burned off. The others must be diluted or washed off with sterile water.

8. *Sterility Checks*

The effectiveness of many of the above sterilization procedures may be checked by including a tube containing spores of *Bacillus subtilis*: Add sterile nutrient broth to the tube containing spores and incubate to see whether or not there is growth. Another sterility check is the use of paper impregnated with a heat-unstable dye which changes color at the sterilization temperature and indicates if the temperature has reached the sterilization range. These are available from most laboratory supply houses.

9. *Sterilizing Inoculating Tools*

A. Sterile Needle or Loop

These may be sterilized in the flame of a bunsen burner or alcohol lamp until red hot and then allowed to cool before touching any other object.

B. L-Shaped Rod or Spatulas

Sterilize by dipping into 95% ethanol for a few minutes and then burning off excess ethanol in flame. Take care not to have flaming alcohol on your hand. Allow the rod to cool without touching any other object before use.

F. Anaerobiosis

Several techniques are available for producing an anaerobic environment, an environment which is low or free of oxygen or one which is reducing. Culture tubes or plates to be incubated anaerobically may be placed in any one of the following anaerobic situations.

1. *Pyrogallol and KOH*

The Bray dish is designed for this procedure. Place 10 ml of 20% aqueous KOH on one side of the bottom of the Bray dish and 4 ml of 40% aqueous pyrogallol on the other side. Streak the surface of a nutrient agar plate to be used for growth of anaerobes. Place agar plate on top of Bray dish. Seal rim of Petri dish and Bray dish with plasticine. Tilt alkali into pyrogallol.

2. Deep Agar Tube

Any agar nutrient medium may be used in a tube filled 4–6 inches with agar medium. The anaerobic organism may be introduced by stabbing agar to the bottom of the tube with a long needle or by shake culture.

3. Reducing Agents

Certain compounds which produce a reducing environment may be added sterilely to various media or just before sterilization. Among these are: sodium thioglycollate, 0.1%; sodium formaldehyde sulfoxylate, 0.1%; sodium thiomalate (mercaptosuccinate), 0.1%; L-cysteine (usually as the hydrochloride; this is very acidic), 0.1%.

4. Evacuation and Replacement

Tubes, flasks, Petri dishes, may be incubated in a sealed container which can be evacuated by a vacuum pump (or less efficiently an aspirator). After evacuating the air the container may be filled to atmospheric pressure with an inert gas such as nitrogen, helium, hydrogen, or carbon dioxide.

5. Sterile Overlay

(a) Light mineral oil may be sterilized by autoclaving and when cool the mineral oil can be poured over the surface of inoculated tubes, flasks, and, less efficiently, Petri dishes to a depth of 1/2–1 inch. This will tend to exclude air.

(b) Paraffin (such as used for canning) may also be sterilized by autoclaving and when cooled, but still molten, can be poured over the surface of inoculated tubes, Petri dishes to a depth of 1/4–1/2 inch. This will tend to exclude air.

6. Hydrogen Generation

Materials

Granulated zinc. Use 3.3 g/liter of container volume.

Dissolve 100 g $CrK(SO_4)_2 \cdot 12\ H_2O$ (Chrome alum) in 480 ml H_2O. Add 120 ml conc. H_2SO_4. When ready to activate zinc add 0.1 g $CuSO_4 \cdot 5\ H_2O$ to solution. Use 20 ml of this solution per liter of container volume.

$CaCO_3$. Use 0.23 g per liter of container volume.

Anaerobic indicator: methylene blue (0.04 g/100 ml in 0.2 *M* phosphate buffer at pH 7.0), 1.0 ml; ascorbic acid (0.5 g/100 ml), 1.0 ml; agar (0.6 g/100 ml), 8.0 ml.

Add indicated volumes of methylene blue and ascorbic acid to melted agar in small tube. Cool until solid and stopper indicator tube to prevent dessication until use.

Any large container not affected by acid which can be made gas-tight by sealing, i.e., McIntosh and Fildes jars, Kelner jars, dessicators, etc.

Procedure

1. Place experimental material (tubes or Petri dishes) and indicator tube in gas-tight container.
2. Add zinc and $CaCO_3$ to reagent dish.
3. Add chromic sulfate plus sulfuric acid solution through a long-stemmed funnel.
4. Evacuate to 80–100 mm mercury.
5. Seal jar and incubate.

Displacement method may be used by following procedure of V. H. Mueller and P. A. Miller, *J. Biol. Chem.* **41**, 301 (1941).

Remember escaping gas is a mixture of hydrogen and air which is flammable. [Reference: J. H. Marshall, *J. Gen. Microbiol.* **22**, 645–8 (1960).]

7. *Miscellany*

The extent of anaerobiosis may be measured by use of a dye which changes color when reduced or oxidized. Methylene blue (0.002%, aqueous) has been used frequently; it is blue when oxidized and colorless when reduced. Another useful and quite nontoxic dye is indigo carmine (sodium indigodisulfonate). Other methods are described in the several microbiology references.

8. *Techniques for Fastidious Anaerobes*

Moore has detailed a number of techniques for the handling and growing of anaerobes which are sensitive to even small amounts of oxygen. [The reader is recommended to his paper: W. E. C. Moore, Techniques for routine culture of fastidious anaerobes. *Intern. J. Systematic Bacteriol.* **16**, 173–190 (1966).]

G. Staining Techniques

1. Bacteria

Bacteria are smeared onto a clean slide with a loop or needle from a liquid medium or mixed in a drop of water before smearing on slide if from a solid medium. When the smear is dry, pass slide quickly through bunsen burner flame 2–3 times to fix bacteria to slide. *Do not heat slide in flame.* Smears may now be stained.

A. GRAM STAIN

Materials

Stain	
Crystal violet	2.0 g
Ammonium oxalate	1.0 g
Water to	100 ml
Burke's iodine	
Iodine	1.0 g
KI	2.0 g
Water to	100 ml
Aqueous safranin	
Safranin O (2.5% in 95% ethanol)	10 ml
Water to	100 ml

Procedure

1. Flood smear with crystal violet for 1 minute; wash stain off.
2. Flood smear with Burke's iodine for 1 minute; wash off iodine solution.
3. Decolorize with 95% ethanol against white background until no color runs off smear; wash alcohol off.
4. Counterstain with aqueous safranin for 30 seconds; wash dye off.
5. Blot smear and then allow to air dry.
6. Examine microscopically under oil immersion.

Observation

Gram-positive bacteria are violet to black. Gram-negative bacteria are red. Some bacteria are gram-variable never appearing all red or violet (or black). Gram-positive bacteria may also become gram-negative with age, change in growth conditions, etc.

B. Ziehl-Neelsen Stain for Acid-Fast Bacteria

Materials

Loeffler's methylene blue	
Methylene blue (saturated solution in ethanol)	30 ml
KOH (0.001% aqueous)	100 ml
Concentrated carbolfuchsin	
Basic fushsin	1.0 g
Absolute ethanol	10 ml
Phenol (5% aqueous)	100 ml

Dissolve dye in alcohol and add phenol solution.

Acid alcohol	
Conc. HCl	3.0 ml
Ethanol (95%) to	100 ml

Procedure

1. Stain smear with carbolfuchsin which is heated until it steams, but does not boil (3-5 minutes); wash well.
2. Decolorize with acid alcohol until color no longer runs out; wash well.
3. Counterstain with Loeffler's methylene blue (4-5 minutes); wash well.
4. Blot and air dry.
5. Examine microscopically under oil immersion.

Observation

Acid-fast bacteria appear red while all other bacteria are blue.

C. Spore Stain

Materials

Malachite green (5% aqueous)

Aqueous safranin	
Safranin O (2.5% in 95% ethanol)	10 ml
Water to	100 ml

Procedure

1. Stain smear in steaming, but not boiling, malachite green (10 minutes); wash smear well.
2. Counterstain with aqueous safranin (1-2 minutes); wash well.
3. Blot and air dry.
4. Examine microscopically under oil immersion.

Observation

Spores appear green. All other cells or parts of cells appear red. See the following references for additional techniques: (1) "Manual of Methods for Pure Culture Study of Bacteria," edited by the Committee on Bacteriological Technique of the American Society of Microbiologists (Society of American Microbiologists). Biotech Publ., Geneva, New York, 1946. (2) V. B. D. Skerman, "A Guide to the Identification of the Genera of Bacteria." Williams & Wilkins Co., Baltimore, Maryland, 1959.

2. *Fungi*

A. Trypan Blue Stain

Materials

Stain	
Trypan blue	0.1-0.5 g
Water	55 ml
Acetic acid, glacial	45 ml

Procedure

1. Stain smears on slides with trypan blue solution. Control staining by examining smear under microscope. Pour off excess stain.
2. Rinse with water.
3. Drain (draw off most of the water with the edge of a piece of filter paper but *do not* allow the smear to dry completely).
4. Mount in Aquamount.
5. Seal with Laktoseal.

Observation

Cell wall stains blue. Cyptoplasm often reddish. Cell structures show well.

B. Media for Mounting and Staining Molds

Materials

Lactophenol	
Lactic acid	100 ml
Phenol	100 g
Glycerol	100 ml
Water	100 ml

Dissolve phenol in water without heat, then add lactic acid and glycerol.

Lactophenol—cotton blue	
Soluble aniline blue (cotton blue), saturated solution	10 ml
Glycerol	10 ml
Water	80 ml

Mix equal parts of lactophenol and cotton blue solution.

Lactophenol—Picric acid	
Phenol	100 g
Lactic Acid	100 ml
Glycerol	100 ml
Picric Acid (saturated solution)	100 ml

Dissolve phenol in picric acid solution without heat, then add lactic acid and glycerol.

Paraffin wax-petroleum jelly mixture (1:1, v/v)

Procedure

1. Immerse mycelium in a drop of any one of the above mounting media.
2. Cover with a cover slip and ring cover slip with paraffin-petroleum jelly to prevent evaporation.

Observation

Examine mycelium for structure. Cotton blue and picric acid media stain hyphae blue and yellow, respectively.

3. *Algae*

A. Azure A Method

Materials

Carnoy's solution	
Acetic acid, glacial	1 vol
Absolute ethanol	3 vol

Azure A

Add 1 drop of thionyl chloride to 10 ml of azure A (0.25%, aqueous) just before use.

1 *N* HCl
Xylene ("xylol")
60°C water bath or incubator

Procedure

1. Smear cells gently on albumin-coated slide; *do not* allow to dry completely.
2. Wash well in distilled water.
3. Hydrolyze in 1 *N* HCl at 60°C (4–10 minutes).
4. Stain slide in azure A dye as described for about 2 hours.
5. Wash dye off in two rapid rinses with distilled water; blot excess water.
6. Place slide in absolute ethanol for 12 hours in freezing compartment of refrigerator to dehydrate.
7. Place slide in absolute ethanol-xylol (1:1, v/v) for a few minutes.
8. Place slide in xylene for a few minutes.
9. Mount in Permount, Canada balsam, etc. with a coverslip. Delicate dyes are preserved best in the new synthetic mounting resins.

Observation

Cell parts appear dark blue on a light blue background.

B. ACETOCARMINE METHOD.

Materials

Modified Johansen fixative

Iodine	0.25 g
KI	1.0 g
Acetic acid, glacial	4.0 ml
Formalin	24.0 ml
Water	400.0 ml

Acetocarmine (National Aniline) in 400 ml glacial acetic acid (45%) for 4 hours. Cool and then filter.

Procedure

1. Gently smear cells on an albumin-coated slide; do not allow to dry completely.
2. Fix in modified Johansen fixative (2–4 hours).
3. Drain off excess fixative and add 2 drops of filtered acetocarmine. Cover with cover slip.
4. Heat slide gently over low flame until vapor arises from stain.
5. Observe immediately under microscope.

Observation

Nuclei red; cytoplasm unstained. [Reference: T. R. Deason, and H. C. Bold, *Univ. Texas Publ. No.* **6022**, pp. 1-71 (1960).]

C. Johansen Method

Materials

Freshwater algae fixative	
Chromic acid (chromium trioxide)	1.0 g
Propionic acid	1.0 ml
Water to	100 ml
Marine algae fixative	
Chromic acid	1.0 g
Propionic acid	1.0 ml
Seawater to	90 ml

Fix for 4-24 hours. Remove fixative by washing in running water or several changes of seawater, respectively.

Mordant	
Ferric ammonium sulfate	2.0 g
Water to	100 ml
Stain	
Hematoxylin	0.5 g
$NaHCO_3$	0.2 g
Water to	100 ml

Heat stain almost to boiling and cool quickly.

Procedure

1. Mordant for 1 hour.
2. Wash well with water.
3. Stain 1-4 hours or until material is black.
4. Differentiate stain with ferric ammonium sulfate (2% in water) or saturated aqueous picric acid until cell parts are clearly seen; wash well with water.
5. Dehydrate freshwater material by treating for 2 hours in each of the following: 10, 20, 35% alcohol in distilled water followed by 2 hours in increasing concentrations of dehydrating medium:

	50%	70%	85%	95%	100%
Water (ml):	50	30	15	0	0
Ethanol (ml);	40	50	50	45	25[a]
Tertiary butyl alcohol (ml):	10	20	35	55	75

[a]Absolute ethanol.

For marine algae first transfer for 1 hour in each of the following series and then continue as for freshwater algae above in 50, 70, 85, 95, and 100% alcohol. The final steps of dehydration and embedding are the same as for vertebrate tissue. [Reference: D. A. Johansen, *in* "Manual of Phycology" (G. M. Smith, ed.), pp. 359-63. Chronica Botanica Co., Waltham, Massachusetts, 1951.]

		10%	15%	20%	30%	40%
Seawater (ml):	90	80	65	50	35	20
Distilled water (ml):	5	10	20	30	35	40
Ethanol (ml):	5	10	15	20	30	40

4. *Protozoa*

A. Fixatives

Schaudinn's Fluid	
$HgCl_2$, saturated aqueous solution	66 ml
Ethanol, 95%	33 ml
Acetic acid, glacial (add immediately before use)	5 ml
Bouin's fluid	
Picric acid, saturated aqueous solution	75 ml
Formalin, concentrated	25 ml
Acetic acid, glacial	5 ml
Hollande's fluid	
Cupric acetate (should be dissolved before adding picric acid)	2.5 g
Picric acid	4.0 g
Distilled water	100 ml
Formalin	10 ml
Acetic acid, glacial	1.5-5 ml

B. STAINING PROCEDURES—PROTARGOL METHOD

Procedure

1. Smear organisms on cover slip smeared with albumin.
2. Fix material on cover slip with Bouin's or Hollande's fluid.
3. Wash with water (2–3 minutes).
4. Transfer smear through 15, 30, 50, 60, 70% ethanol series. Change ethanol until color no longer appears.
5. Transfer smears back down ethanol series to water. Transfer to $KMnO_4$ solution (0.25% aqueous) for 5–10 minutes.
6. Wash with distilled water for several minutes. Smears now appear dark brown.
7. Place smears in 2–2.5% oxalic acid (5–10 minutes). Brown color is bleached.
8. Wash smears well.
9. Dissolve Protargol S just before use in distilled water to make a 1.5% solution. Add 0.5 g copper as 18- or 20-gauge wire in bottom of container. Leave smears in activated Protargol S for 18–36 hours at 30°C.
10. Rinse smears in distilled water.
11. Reduce smears in following solution for 5–10 minutes.

$NaSO_3$, anhydrous	5.0 g
Hydroquinone	1.0 g
Water	100 ml

Dissolve $NaSO_3$ completely before adding hydroquinone. Smear will turn darker brown.

12. Wash smears well.
13. Transfer smears to gold chloride solution (1%) for about 5 minutes or until smears become grayish in color.
14. Rinse smear in distilled water twice.
15. Transfer smear to oxalic acid (2%, aqueous) for 5–10 minutes; smears become purple.
16. Wash well.
17. Place smear in sodium thiosulfate (5%, aqueous).
18. Wash well for at least 10–20 minutes.
19. Dehydrate through ethanol series, clear in toluene, and mount in balsam or other permanent mounting medium. [(See References for additional information on staining, etc.) Reference: A. E. Galigher, and E. N. Kozloff, "Essentials of Practical Microtechnique." Lea & Febiger, Philadelphia, Pennsylvania, 1964.]

H. Preparation of Silica Gel Plates*

Materials

20% aqueous *o*-phosphoric acid (ACS, 85%, from Fisher Scientific Co.).

Dissolve 10 g of powdered silica gel (100–200 mesh; Will Scientific Co.) or silicic acid (reagent grade, J. T. Baker Chemical Co.) in 100 ml of 7% (w/v) aqueous KOH by heating.

Appropriate nutrient medium for microorganisms in double strength.

Procedure

1. Sterilize 4.0 ml-portions of the phosphoric acid, 10.0 ml-portions of the potassium silicate, and 10.0 ml of the double-strength nutrient medium in individual tubes for each solution.
2. When tubes are sterile, mix contents of tubes into Petri plates in the following order: first mix 10 ml potassium silicate solution with 10 ml of nutrient medium. Next add 4.0 ml of phosphoric acid and mix.
3. Allow plates to harden at room temperature (approximately 15 minutes). The silica gel will undergo *syneresis* (removal of water from the gel).
4. Remove excess liquid with a sterile pipette or pour off aseptically.
5. Streak plates in the usual manner and incubate in the usual manner in a moist atmosphere.
6. Plates dry easily. To prevent drying and cracking of the silica gel, evaporization may be minimized by sealing the edge of the plate with Parafilm or placing the entire plate in a clear or translucent plastic sandwich bag and sealing.

*These may be used in place of agar or to replace agar in the purification of agar liquifiers.

VI

Microbial Nutrition

A. General Considerations

Microorganisms, like all organisms, draw on external chemicals for synthesis of their protoplasm and for energy for survival, growth, and multiplication. This energy, usually intracellularly in the form of ATP (adenosine triphosphate), is generated by one of three methods: (*i*) *Photophosphorylation*—synthesis of ATP by converting light energy into the energy stored in the ATP molecule by the process found in photosynthetic bacteria, algae, phytoflagellates, and higher plants; (*ii*) *Lithooxidative phosphorylation*—removal of electrons from inorganic molecules such as H_2S, S, $Na_2S_2O_3$ NH_3, NO_2^-, H_2, $Fe(HCO_3)_2$, or CO coupled with the synthesis of ATP, carried out by chemoautotrophic organisms, mainly bacteria; (*iii*) *Organooxidative phosphorylation*—electrons and protons (H^+) are removed from organic molecules coupled with the synthesis of ATP in heterotrophic organisms, most bacteria, fungi, and all animals. (See Table VI-1 and Fig. VI-1 for examples of these organic molecules.) To a greater or lesser extent, this process contributes to the ATP store of all chemoautotrophic and photosynthetic organisms, depending on their permeability to organic molecules and the availability of such foodstuffs.

TABLE VI-1

SOME CARBON AND ENERGY SOURCES

Carbohydrates, sugar, and sugar alcohols	TCA-cycle acids
Arabinose	Acetic
Cellulose	Aconitic
Dextrin	α-Ketoglutaric
Fructose	Citric
Galactose	Fumaric
Gentianose	Malic
Glucose	Oxaloacetic
Glycerol, dulcitol, sorbitol	Succinic
Glycogen	Compounds leading to TCA cycle
Inulin	Amino acids
Lactose	Alanine
Maltose	Arginine
Mannitol, adonitol	Aspartic acid, asparagine
Melezitose	Glutamic acid, glutamine
Melibiose	Glycine
Raffinose	Serine
Ribose	Fatty acids
Rhamnose	Butyric acid
Starch	Propionic acid
Sucrose	
Trehalose	
Xylose	

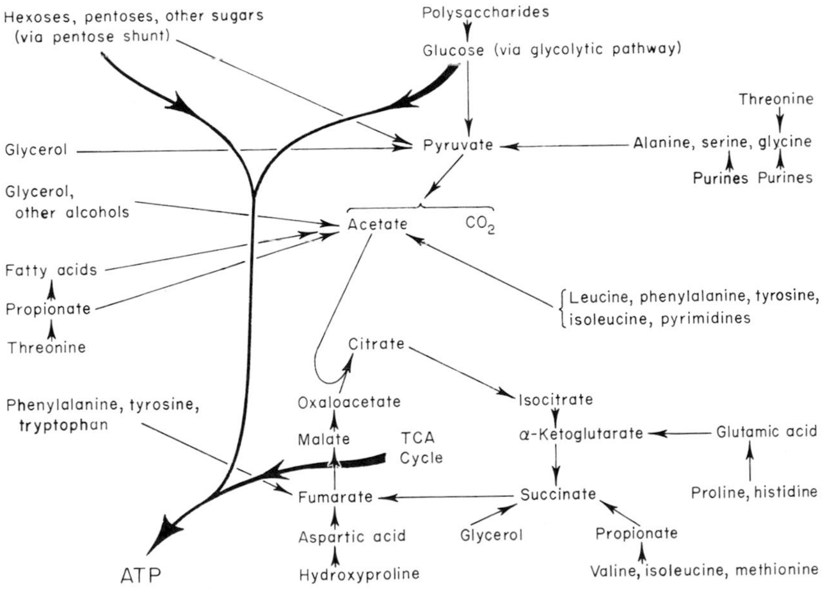

FIG. VI-1. Some alternate pathways for supplying carbon and energy.

Besides an energy source, all organisms need major elements as carbon, oxygen, hydrogen, nitrogen, phosphorus, sulfur, potassium, and magnesium for the synthesis of major cellular structures and, as noted earlier, minor or trace elements: sodium, calcium, manganese, zinc, iron, copper, molybdenum, vanadium, boron, cobalt, chlorine, iodine, silicon, for the synthesis of coenzymes, enzymes, vital molecules, specialized structures, salt formation, osmotic regulation, etc. (see Table VI-2). For autotrophic microorganisms, which can synthesize all or most of their organic molecules, these elements may be supplied as inorganic compounds. Organic molecules must be supplied to the heterotrophs and those autotrophic microorganisms unable to synthesize specific organic molecules, notably vitamins or other microbial "growth factors" (see Table VI-2). The elements required for life fall into the following ranges (supplied per 100 ml of nutrient medium). *Major elements* must be supplied in 0.01 g to 1 g amounts; *minor elements* in 0.1 to 10.0 mg amounts; and *trace elements* 0.1 to 10 μg amounts or as contaminants in the major and minor elements.

In autotrophic microorganisms, e.g., most photosynthetic and chemoautotrophic bacteria, all nutrients are usually supplied as inorganic salts except for the occasional requirement for one or more vitamins. In many heterotrophic microorganisms the major elements and energy sources may be selected from a wide assortment of organic molecules (Table VI-3) while the minor and trace elements are supplied as inorganic compounds or as contaminants in other compounds [or vitamin B_{12}].

B. Organic Nutrition

Many heterotrophic and some autotrophic microorganisms will not grow or multiply unless given small quantities of certain organic molecules that they cannot synthesize for themselves. Usually these molecules are vitamins, but they may also be molecules that are required for the synthesis of macromolecules like nucleic acids and proteins. Microorganisms are known which also require one or more of the following: purines, pyrimidines, amino acids, unsaturated fatty acids, sterols, and heme. These compounds unlike the vitamins are usually required in so high a concentration that multiplication of the organism can be immediately arrested if they are absent from the medium. On the other hand, vitamins (see Table VI-3) may be required immediately, or the organisms used for inoculation may contain enough to give a misleading impression that the vitamin is not

TABLE VI-2
ELEMENTS REQUIRED FOR LIFE

Element	Symbol	Form usually supplied	Biological role
Major			
Carbon	C	CO_2, organic compounds	Structure, organic compounds
Oxygen	O	CO_2, O_2, H_2O, organic compounds	Structure, organic compounds
Hydrogen	H	H_2O, organic compounds	Structure, organic compounds
Nitrogen	N	NH_4^+, NO_3, organic compounds	Structure, organic compounds
Minor			
Phosphorus	P	PO_4, organic compounds	ATP, nucleic acids, phospholipids
Sulfur	S	SO_4, H_2S, organic compounds	Amino acids, vitamins, sulfolipids
Sodium	Na	Inorganic or organic salts	Salts, osmosis, regulation of permeability
Potassium	K	Inorganic or perhaps organic compounds	Salts, osmosis, enzyme function
Calcium	Ca	Inorganic or organic salts	Salts, osmosis, membranes, enzyme function
Magnesium	Mg	Inorganic compounds	Enzyme cofactors, chlorophyll
Trace			
Manganese	Mn	Inorganic compounds	Enzymes
Zinc	Zn	Inorganic compounds	Enzymes
Iron	Fe	Inorganic compounds	Enzymes
Copper	Cu	Inorganic compounds	Enzymes
Molybdenum	Mo	Inorganic compounds	Enzymes
Cobalt	Co	Inorganic compounds	Enzymes, vitamin B_{12}
Vanadium	V	Inorganic compounds	Enzymes of nitrogen fixation
Boron	B	Inorganic compounds	Required by plants
Silicon	Si	Inorganic compounds	Skeleton of radiolaria, diatoms
Chloride	Cl	Inorganic compounds	Enzymes
Iodide	I	Inorganic compounds	Thyroxine
Selenium	Se	Inorganic compounds	May be required by animals

TABLE VI-3
VITAMINS AND GROWTH FACTORS

Compound	Other nutritional forms	Required by
p-Aminobenzoic acid	Folic acid	Microorganisms
Folic acid	Folinic acid	Microorganisms; man
Biotin		Microorganisms; man
Nicotinic acid	Nicotinamide	Microorganisms; man
Pantothenic acid	Pantethein	Microorganisms; man
Riboflavine	Riboflavine monophosphate	Microorganisms; man
Thiamine		Microorganisms; man
Pyridoxal	Pyridoxine Pyridoxamine Pyridoxal phosphate	Microorganisms; man
Vitamin B_{12}		Microorganisms; man
Thioctic acid (lipoic acid)		Microorganisms, notably ciliates
Heme	Hemin	Microorganisms, notably certain flagellates and yeasts
Carnitine		Insects
Inositol		Yeast
Menadione; Phthiocol (vitamin K)		Acid-fast bacteria
Cholesterol; other steroids		Insects; some protozoa
Sideramines		
Ferrichrome	Various lipid-soluble chelating agents	Microorganisms
Coprogen		Microorganisms
Terregens factor		
Choline		Microorganisms
Putrescine	Spermidine, spermine	Microorganisms
Bifidus factor	Glycosides of *N*-acetyl-D-glucosamine	Microorganisms, certain lactic acid bacteria
Biopterin	2-Amino-4-hydroxy-6-substituted pteridines	Microorganisms
Glutamine		Microorganisms
Asparagine		Microorganisms; some insects
Oleic acid	Other unsaturated fatty acids	Microorganisms

required. This type of sensitive response can only be elicited after serial incubations of cells in a medium lacking the vitamin; only after near-complete depletion from cell stores will this type of vitamin requirement emerge. The unwary should be warned that laboratory dust, finger grease, or "chemically pure" organic molecules of biological origin may contain enough vitamin.

Certain organic molecules, while not absolutely required for the growth and multiplication of a particular microorganism, may, if supplied in the nutrient medium, yield more cells or shorten the time needed to reach maximal cell number. These stimulatory nutrients may render a metabolic sequence more efficient or, by their use, the cells are spared the need to synthesize them and the energy and raw materials are then used for other needed syntheses, permitting more efficient metabolism.

C. Identifying Nutrients

The procedures for identifying required or stimulatory nutrients are diverse and numerous; those included here are not the only ones that can be used. The ones outlined here are those found most satisfactory in the author's laboratory.

1. General Conditions

The microorganism whose nutrition is to be analyzed must, if at all possible, first be grown in laboratory culture in any nutrient solution, or on agar, under conditions which permit multiplication. It must be free of contaminating organisms. While an occasional growth factor can be identified even from mixed cultures, it is needlessly difficult to identify the majority of nutrients which may be exchanged between members of a mixed population. If possible, clones (populations consisting of the offspring of a single microorganism) should be used; this will avoid problems of genetic recombination and shifting nutrient requirements that might result from heterozygous populations. It will not, however, eliminate mutations which may express themselves immediately in haploid cells, i.e., bacteria; this is not a serious problem in studying microbial nutrition but it should be kept in mind if nutrient requirements seem to change with time in a clone.

The physiological conditions of growth, particularly light, presence of oxygen, temperature, and pH, may affect nutrition; therefore, these must be controlled as rigorously as possible. Many photosynthetic microorganisms can use light and CO_2 as sole source of carbon and energy; others need vitamins and other growth factors besides; some may grow in minimal media in the light but require more elaborate nutrients in the dark; still others may have minimal requirements in the light in the absence of oxygen and additional nutrient needs in the light in the presence of oxygen. Many microorganisms require additional nutrients and drastic shifts in the proportions of nutrients for growth at temperature above or below those at which their nutritional requirements have been established with conventional culture media. pH will modify the ability of polar nutrients to pass through the cell membrane. Therefore, nutrient requirements may be modified by marked changes in pH even if they do not prevent multiplication.

Nutritional requirements may seemingly be modified by changes in the purity of organic nutrients. Many organic compounds, especially those isolated from natural materials, often contain varying amounts of contaminating organic or inorganic molecules. Inorganic salts are usually contaminated with sufficient trace elements to satisfy a good part or even all of an organism's needs under ordinary conditions. The quality of commercial organic compounds is steadily improving. If the purity of an organic compound is in question its homogeneity may be tested by chromatographic or other means.

2. *Inorganic Nutrients*

Every microorganism needs major and minor elements. The major elements with the exception of carbon and nitrogen are usually supplied as inorganic molecules. The elements are supplied as compounds shown in Table VI-4. Examples of marine and freshwater media are shown in Table VI-5. Each new organism may require modifications in the concentration of one or more of its inorganic nutrients. This can be shown by removing the specific element and titrating growth against increasing concentrations of the element. The trace elements needed may be more difficult to identify since they often contaminate the major elements. Several techniques are, however, available. One involves the addition of trace element mixes (Tables VI-6 and 7) and successive elimination of each element to see if it is required. Another method involves titrating a particular element against fixed amounts of the other elements and a chelator.

3. *Organic Nutrients.*

Organic nutrients may be identified by trying to grow the microorganism on a progressive series of nutrients starting with ill-defined nutrients "gunks," e.g., soil extract, trypticase (a tryptic digest of casein), liver extracts, yeast autolyzate, beef extract, peptones, serum, blood, etc., and then by replacing or reducing the concentration of "gunk" required by replacing it with better defined organic mixtures such as "complete supplement," (see Table VI-8) vitamin-free acid hydrolyzate of casein (e.g., Hycase), gelatin hydrolyzate, hydrolyzed

TABLE VI-4
COMPOUNDS USED TO SUPPLY MAJOR AND MINOR ELEMENTS

Element	Compounds	Conc. range[a]
C	CO_2, CO_3^{2-}, organic molecules	g
O	O_2, H_2O, organic molecules	g
H	H_2O, organic molecules	g
N	N_2, NH_4, NO_3^-, NO_2^-, amino acids, purines, pyrimidines	g
Na	Several inorganic salts, i.e., NaCl, Na_2SO_4, Na_3PO_4	g
K	Several inorganic salts, i.e., KCl, K_2SO_4, K_3PO_4	g
Ca	Several inorganic salts, i.e., $CaCO_3$, Ca (as chloride)	g
P	Several inorganic salts, Na or K phosphates, Na_2 glycerophosphate · 5 H_2O	g
S	Several inorganic salts, $MgSO_4$ · 7 H_2O, amino acids	g
Mg	Several inorganic salts, CO_3^{2-}, SO_4^{2-} or CL^- salts	g
Fe	$FeCl_3$, $Fe(NH_4)_2SO_4$, ferric citrate	mg
Zn	SO_4^{2-} or Cl^- salts	mg
Mn	SO_4^{2-} or Cl^- salts	mg
Cu	SO_4^{2-} or Cl^- salts	μg
Co	Vitamin B_{12}, SO_4^{2-} or Cl^- salts	μg
B	H_3BO_3	mg
Mo	Na or NH_4 molybdate salts	μg
V	Na_3VO_4 · 16 H_2O	μg
Sr	SO_4^{2-} or Cl^- salts	μg
Al	SO_4^{2-} or Cl^- salts	μg
Rb	SO_4^{2-} or Cl^- salts	μg
Li	SO_4^{2-} or Cl^- salts	μg
Cl	as Na^+, K^+, $^+Ca^{2+}$, of NH_4^+ salts	g
I	as Na^+, K^+, Ca^{2+}, or NH_4^+ salts	μg
Br	as Na^+, K^+, Ca^{2+}, or NH_4^+ salts	mg
Si	Na_3SiO_3 · 9 H_2O	mg

[a]per 100 ml of medium.

DNA and RNA. These organic mixtures in turn are replaced by known mixtures, i.e., amino acids, purines, pyrimidines, their nucleosides and nucleotides, vitamins, and minerals. When the required groups of organic compounds have been identified, the individual compound may be identified by subdividing the group of related organic compounds into smaller groups, and these in turn to individual compounds, until each required or stimulatory compound has been identified. The larger and smaller groups are listed in Tables VI-9 to VI-16.

TABLE VI-5
COMPARISON OF MAJOR AND MINOR ELEMENT REQUIREMENTS FOR MULTIPLICATION[a]

	Freshwater			Marine		
Nutrient	*Ochromones danica*[(1)e]	*Phacus pyrum*[(2)e]	*Micrococcus sodonensis*[(3)e]	*Gyrodinium* sp.[(4)e]	*Stichococcus* sp.[(4)e]	*Labyrinthula* sp.[(4)d,e]
NH_4Cl	0.05 g		0.05 g			
$(NH_4)_2SO_4$		0.05 g				
KCl		0.003 g		0.03 g		
NaCl				2.4 g	2.3 g	2.5 g
$MgCl_2 \cdot 6\,H_2O$				0.3 g	1.1 g	
$MgSO_4 \cdot 7\,H_2O$	0.1 g	0.03 g	0.2 g	0.03 g		0.5 g
Na_2SO_4					0.4 g	
$MgCO_3$	0.04 g					
$NaHCO_3$					0.2 g	
KNO_3			0.01 g	0.01 g		
$Ca(NO_3)_2 \cdot 4\,H_2O$					0.01 g	
$CaCO_3$	0.005 g		0.02 g	0.02 g		0.025 g
K_2HPO_4					0.002 g	0.01 g
KH_2PO_4	0.03 g	[b]	[b]	0.002 g		
Fe	0.2 mg	100 μg	0.4 mg	0.3 mg	0.05 mg	0.2 mg
Zn	0.1 mg	1.0 μg	1.0 mg	0.4 mg	0.03 mg	2.0 mg
Mn	0.05 mg	9.0 μg	0.8 mg	1.0 mg	0.01 mg	2.0 mg

Cu	0.008 mg	0.03 μg	0.1 mg	0.3 μg	0.01 mg	
Co	0.01 mg	0.3 μg	0.1 mg	3.0 μg	[c]	[c]
B	0.01 mg		0.4 mg	0.2 mg	0.01 mg	
Mo	0.005 mg	0.2 μg	0.08 mg	0.05 mg	0.01 mg	
V	0.001 mg					
Ca		2.0 mg	1.0 mg			
Sr				0.65 mg		
Al				0.025 mg		
Rb				0.01 mg		
Li				0.005 mg		
I				2.5 μg		
Br				3.25 mg		
pH	5.0	5.5	7.0	7.5	8.0	8.0-8.2

[a]Concentration of elements per 100 ml of medium (in H_2O).
[b]Supplied as Na_2 glycerophosphate · 5 H_2O.
[c]Supplied as vitamin B_{12}.
[d]Supplied with gelatin hydrolyzate and other vitamins.

[e]REFERENCES:

(1) S. Aaronson, and S. Scher, *J. Protozool.* **7**, 156–158 (1960).
(2) L. Provasoli, and I. J. Pintner, *In* "The Ecology of Algae" (C. A. Tryon, Jr. and R. T. Hartman, eds.), pp. 84-86. Univ. of Pittsburgh Press, Pittsburgh, Pennsylvania, 1960.
(3) S. Aaronson, *J. Bacteriol.* **69**, 67–70 (1955).
(4) L. Provasoli, J. J. A. McLaughlin, and M. R. Droop, *Arch. Mikrobiol.* **25**, 392–428 (1957).

TABLE VI-6
MARINE TRACE ELEMENT MIXTURES[a]

TM2 trace metals[b]		
Na_2 EDTA	2.0	mg
Fe (as Cl^-)	0.01	mg
Mn (as SO_4^{2-})	0.065	mg
Zn (as SO_4^{2-})	0.23	mg
Mo (as $NaMoO_4$)	0.02	mg
Co (as SO_4^{2-})	0.00063	mg
Cu (as SO_4^{2-})	0.00013	mg
ASP trace metals[b]		
Na_2 EDTA	1.0	mg
Fe (as Cl^-)	0.01	mg
Zn (as Cl^-)	5.0	μg
Mn (as Cl^-)	0.04	μg
Co (as Cl^-)	0.1	μg
Cu (as Cl^-)	0.04	μg
B (as H_3BO_3)	0.2	mg
RC trace metals[b]		
Na_2 EDTA	1.0	mg
Fe (as Cl^-)	0.01	mg
Zn (as Cl^-)	5.0	μg
Mn (as Cl^-)	0.04	mg
Co (as Cl^-)	0.1	μg
Cu (as Cl^-)	0.04	μg
B (as H_3BO_3)	0.2	mg
Br (as K^+)	6.5	mg
Sr (as Cl^-)	1.3	mg
Al (as Cl^-)	0.05	mg
Rb (as Cl^-)	0.02	mg
Li (as Cl^-)	0.01	mg
I (as K^+)	0.005	mg

[a]After L. Provasoli, J. J. A. McLaughlin and M. R. Droop *Arch. Mikrobiol.* **25**, 392–428 (1957).
[b]Use these amounts per 100 ml of medium.

Organic nutrients include carbon and energy sources, such as those listed in Table VI-1, which are required in large quantity; organic compounds are required in somewhat lesser amounts for the synthesis of macromolecules or cell structures; and organic compounds such as vitamins (Tables VI-9–11) are required in small quantity for the synthesis of coenzymes and prosthetic groups of enzymes. Amino acids compose proteins. Many microorganisms have been described needing one or more amino acids for growth and multiplication (see Tables VI-15 and 16). Occasionally there have been reports of microorganisms requiring peptides. Further study has shown that these microbes secrete enzymes which destroy required free amino acids; the peptides are less vulnerable and, therefore, seem to be required. Other complex interrelationships affecting amino acid requirements have also been noted; among them are antagonisms between two amino acids and variation in amino acid transport mechanisms across cell membranes. The occasional reports of protein requirements have not been confirmed or the proteins have been shown to be nonspecific detoxifiers.

Purines, pyrimidines, their ribosides, deoxyribosides, ribotides, and deoxyribotides are also often required for or stimulate microbial multiplication (see Table VI-12-14). Sometimes with or without

TABLE VI-7

FRESHWATER TRACE ELEMENT MIXTURES[a]

Freshwater algae A_5 trace elements[b]	
H_3BO_3	0.286 g
$MnCl_2 \cdot 4\ H_2O$	0.181 g
$ZnSO_4 \cdot 7\ H_2O$	0.022 g
MoO_3 (85%)	0.00177 g
$CuSO_4 \cdot 5\ H_2O$	0.0079 g
Water to	100 ml
H_5 trace elements[b]	
$ZnSO_4 \cdot 7\ H_2O$	0.882 g
$MnCl_2 \cdot 4\ H_2O$	0.144 g
MoO_3 (85%)	0.071 g
$CuSO_4 \cdot 5\ H_2O$	0.157 g
$Co(NO_3) \cdot 6\ H_2O$	0.049 g
Water to	100 ml

[a]After W. A. Kratz, and J. Myers, *J. Botany* 42, 282–7 (1955).
[b]Use at 1.0 ml per liter of medium.

TABLE VI-8

COMPLETE SUPPLEMENT[a,b]

Acid-hydrolyzed DNA (2 hours 6 NHCl, 120°C)	0.02 g
Alkaline-hydrolyzed RNA (4 hours, conc. NH_4OH, 180°C)	0.02 g
DL-Methionine	0.006 g
Hycase	0.4 g
L-Tryptophan	0.005 g
Simple vitamin mixture	1.0 ml
Water to	100 ml

[a]In water.
[b]Use at 5.0 ml per 100 ml of medium.

TABLE VI-9
COMPLETE VITAMIN MIXTURE[a]

	Amount per ml
N-Acetylglucosamine	0.3 mg
p-Aminobenzoic acid	0.01 mg
Vitamin B_{12}	0.4 μg
Biotin	1.0 μg
Betaine	0.1 mg
Ca pantothenate	0.1 mg
DL-Carnitine · HCl	0.04 mg
Choline H_2 citrate	1.6 mg
Cystamine · 2 HCl (2,2′-dithiobis-ethylamine · 2 HCl)	0.01 mg
Ferulic acid (4-hydroxy-3-methoxycinnamic acid)	0.01 mg
Inositol	1.0 mg
Nicotinic acid	0.1 mg
Putrescine · 2 HCl	0.02 mg
Pyridoxal ethylacetal	0.01 mg
Pyridoxamine · 2 HCl	0.02 mg
Pyridoxine	0.02 mg
Riboflavine	0.01 mg
Na riboflavine PO_4 · 2 H_2O	0.01 mg
Spermidine PO_4 · 3 H_2O	0.01 mg
Thiamine NO_3	0.04 mg
p-Hydroxybenzoic acid	0.01 mg
Folic acid	0.05 mg
DL-Thioctic acid	0.005 mg

[a]Use at 1.0 ml/100 ml medium.

TABLE VI-10
SIMPLE VITAMIN MIXTURE[a]

	Amount per ml
Thiamine · HCl	0.06 mg
Nicotinic acid	0.1 mg
Ca pantothenate	0.1 mg
Choline H_2 citrate	1.6 mg
Na riboflavine phosphate · 2 H_2O	0.01 mg
Pyridoxamine · 2 HCl	0.02 mg
Pyridoxal · HCl	0.02 mg
Inositol	1.0 mg
Vitamin B_{12}	0.4 μg
Biotin	1.0 μg
DL-Thioctic acid	4.0 μg
p-Aminobenzoic acid	0.01 mg
Folic acid	0.04 mg

[a]Use at 1.0 ml/100 ml of medium.

TABLE VI-11

VITAMIN SUBMIXTURES[a]

	Amount per ml
I	
Thiamine · HCl	0.2 mg
Nicotinic acid	0.2 mg
Ca pantothenate	0.2 mg
Biotin	1.0 μg
Vitamin B_{12}	0.5 μg
II	
Pyridoxamine · 2 HCl	0.025 mg
Na riboflavine phosphate · 2 H_2O	6.0 μg
Choline H_2 citrate	1.0 mg
Inositol	1.0 mg
Putrescine · 2 HCl	0.04 mg
III	
p-Aminobenzoic acid	0.01 mg
p-Hydroxybenzoic acid	0.01 mg
DL-Thioctic acid	0.004 mg
Folic acid	0.001 mg
Folinic acid	0.2 μg

[a]Use at 1.0 ml per 100 ml medium.

TABLE VI-12

PURINE-PYRIMIDINE MIXTURE[a]

	Amount (mg/ml)
Adenine	0.5
Adenosine	1.0
Adenylic acid (2′, 3′)	2.0
Cytidylic acid	2.0
Deoxyguanosine	1.0
Guanine	0.4
Guanosine · H_2O	1.0
Hypoxanthine	0.4
Na inosinate	4.0
Na_2 guanylate · H_2O	2.0
Orotic acid	0.4
Thymidine	1.0
Thymine	0.5
Uracil	0.5
Uridine	1.0

[a]Use at 1.0 ml/100 ml of medium

TABLE VI-13
PURINE MIX[a]

	Amount (mg/ml)
Adenine	0.5
Adenosine	1.0
Adenylic acid (2′, 3′)	1.2
Hypoxanthine	0.5
Guanine · HCl	0.5
Guanosine	1.0
Guanylic acid	1.2
Inosine	1.0
Xanthine	0.5
Xanthosine	1.0

[a]Use at 1.0 ml/100 ml of medium.

methionine one or more of these compounds may replace vitamin B_{12} or folic acid. The biotin requirement may be spared in some bacteria by oleic acid (or other unsaturated fatty acids) and aspartic acid.

4. *Chelation*

Mixed metal solutions which tend to precipitate, especially at higher pH or in the presence of Ca^{2+}, Mg^{2+}, and phosphates, may be kept in solution by lowering the concentration of these ions and the pH as much as is consonant with multiplication. Phosphate may be provided as glycerophosphate salts which are more soluble and provide a phosphate source for many microorganisms. An effective procedure is to provide a compound which attracts and holds metal ions in a soluble form by chelation, a form of noncovalent bonding which is dependent on local electron sharing, and involves a ring formation with resonance stabilization much as in the stabilization of aromatic rings. Some chelators, e.g., citrate or histidine, are metabolizable and may therefore disappear with multiplication; others are not metabolized, e.g., ethylenediaminetetraacetic acid (EDTA, Versene). Addition of a chelator requires titration of the metals to determine the amounts which must be supplied in the presence of a fixed amount of chelator. Useful chelators are listed in Table VI-17.

5. *pH and Buffers*

Microorganisms are known which can grow and multiply at extreme ranges of pH. Most microorganisms, however, grow only

TABLE VI-14
PYRIMIDINE MIX[a]

	Amount (mg/ml)
Cytidine	1.0
Cytosine	0.5
Cytidylic acid	1.2
Orotic acid	0.4
Thymine	0.5
Thymidine	1.0
Uracil	0.5
Uridine	1.0

[a]Use at 1.0 ml/100 ml of medium.

within a limited pH range. Most marine forms do best at pH 7–9; freshwater forms range from a pH of 8.0 to a pH of 1.5–3.0 for the acid flagellates. Most organisms tend to change the pH of their nutrient medium as they grow by removing nutrients and/or secreting metabolites. In either case, buffers usually must be added to prevent

TABLE VI-15
AMINO ACIDS MIXTURE[a]

	Amount (%)	For 500 ml (g)
DL-Alanine	0.04	4.0
L-Arginine (free base)	0.03	3.0
DL-Aspartic acid	0.05	5.0
L-Glutamic acid	0.1	10.0
Glycine	0.05	5.0
L-Histidine (free base)	0.02	2.0
DL-Isoleucine	0.005	0.5
DL-Leucine	0.005	0.5
DL-Lysine · HCl	0.045	4.5
DL-Methionine	0.006	0.6
DL-Phenylalanine	0.004	0.4
L-Proline	0.004	0.4
DL-Serine	0.01	1.0
DL-Threonine	0.01	1.0
DL-Tryptophan	0.005	0.5
L-Tyrosine	0.004	0.4
DL-Valine	0.005	0.5
pH 6.5		

[a]Use at 5.0 ml/100 ml of medium.

TABLE VI-16
AMINO ACID SUBMIXTURES

	Amount (%)	
I[a]		120 ml
DL-Aspartic acid	0.05	×24=1.2 g
L-Glutamic acid	0.08	×24=1.92 g
DL-Alanine	0.05	×24=1.2 g
Glycine	0.05	×24=1.2 g
II[a]		120 ml
L-Arginine (free base)	0.04	×24=0.96 g
L-Histidine (free base)	0.04	×24=0.96 g
DL-Lysine · HCl	0.05	×24=1.2 g
III[b]		250 ml
DL-Isoleucine	0.005	×25=0.125 g
DL-Phenylalanine	0.004	×25=0.1 g
DL-Leucine	0.005	×25=0.125 g
L-Tyrosine	0.004	×25=0.1 g
IV[a]		120 ml
DL-Methionine	0.005	×24=0.12 g
DL-Threonine	0.01	×24=0.24 g
DL-Tryptophan	0.005	×24=0.096 g
V[a]		120 ml
L-Proline	0.004	×24=0.096 g
DL-Serine	0.01	×24=0.24 g
DL-Valine	0.005	×24=0.12 g

[a]Use at 5.0/ml/100 ml of medium.
[b]Use at 10.0 ml/100 ml of medium.

marked pH changes. Table VI-18 indicates some buffers which may be used to prevent these changes. Autoclaving may drastically change the pH of complex solutions even in the presence of buffers, therefore, the pH should be checked before and after sterilization. Buffer capacity also varies with marked temperature changes and pH should be determined at the incubation temperature.

6. *Preservatives*

Many organic media or solutions may be stored at room temperature or in the refrigerator for long periods without deterioration if prevented from becoming a nutrient medium for contaminating

TABLE VI-17
CHELATORS

Compound	pH for Best use
Ethylenediaminetetraacetic acid (EDTA)	6.0–7.5
Nitrilotriacetic acid	Below 6.0
Citric acid	Below 7.0
DL-Malic acid	6.0–7.0
Ethylenediamine(*o*-hydroxyphenyl) acetic acid	6.0–7.5
1,2-Diaminocyclohexane tetraacetic acid	6.0–7.5
Diethylenetriaminepentaacetic acid	6.0–7.5
Ethylene glycol bis(aminoethylether) tetraacetic acid	7.0–7.5
Dihydroxyethylglycine	7.0–7.5

TABLE VI-18
BUFFERS[a]

Compound	Concentration	Best pH for use
Tris(hydroxymethyl)-aminomethane	0.01–0.1%	7.5–8.5
Triethanolamine	0.03%	7–8
L-Histidine[b]	0.03%	6–7
Glycylglycine	0.08%	7–8.2
Succinic acid	0.03%	6–7
Malic acid[b]	0.03%	6–7
N,N-bis(2-hydroxyethyl)-glycine	0.1 *M*	8–9
N-tris(hydroxymethyl)-glycine	0.1 *M*	7.5–8.5
N-2-hydroxyethyl-piperazine-*N*′-2-ethanesulfonic acid	0.1 *M*	7–8
N-tris(hydroxymethyl)-methyl-2-amino-ethanesulfonic acid	0.1 *M*	7–8
2-(*N*-morpholine)ethane-sulfonic acid	0.1 *M*	6–7

[a]See S. P. Colowick and N. O. Kaplan (Eds.), "Methods in Enzymology," Vol. I, pp. 138–146, Academic Press, New York.

[b]*Note:* these are also chelating agents, but much less strongly so than EDTA-type chelators.

organisms which can multiply, albeit slowly, at refrigerator temperatures. The major features of compounds that prevent contamination are (1) that they are potent inhibitors of microbial, especially of mold, growth and (2) that they be relatively nonvolatile at room and refrigerator temperatures but easily and completely volatile at autoclaving temperature. The latter is necessary if the nutrient compounds or media are to be useful. A number of mixtures are available. The mixture used in our laboratory is that developed by S.H. Hutner and C. Bjerknes consisting of the following organic solvents in parts by volume: chlorobenzene-1,2-dichloroethane-1-chlorobutane, 1:1:2, respectively. A few drops of this mixture in a large volume prevents microbial growth for months in a sealed container even at room temperature. Renewal of the preservative is easily made by adding a few drops to the solution if the volatile preservative has evaporated. Another useful preservative is carbon tetrachloride-toluene, 1:1 p.p.v.

VII

Procedures for the Enrichment and/or Isolation of Microorganisms

I. General Directions

Water

The water referred to throughout this book is distilled water unless otherwise described. Tap water may replace distilled water in some procedures if it is allowed to stand for several weeks, boiled, or aerated to remove most of the chlorine.

Aged Seawater

Seawater is stored in the dark at room temperature for several weeks, months, or years to mineralize its organic constituents.

Seawater

Seawater varies somewhat from place to place and also whether it is taken from inshore or offshore or from polluted waters. Those unable to obtain seawater or unpolluted seawater may use one of the several artificial seawaters listed elsewhere in this book.

Microbiological Media and Chemicals

When possible all chemicals are reagent grade or better. Media and chemicals may be purchased from scientific supply houses. A list of sources appears in the annual "Guide to Scientific Instruments" in the Journal, *Science.*

Glassware and Equipment

This equipment may be purchased from scientific supply houses. A list of these supply houses appears in the annual "Guide to Scientific Instruments" in *Science.*

II. Bacteriological Techniques

A. *General*

1. Enrichment for Gram-Positive Bacteria with Sodium Azide

Materials

Nutrient agar, Petri dishes containing sodium azide (0.03 g/100 ml), or any other nutrient medium containing sodium azide (0.03 g/100 ml).

Procedure

1. Streak surface of azide plates with soil or water sample.
2. Incubate for several days at 25°–30°C.

Observation

Azide inhibits the growth of most gram-negative bacteria, therefore, the predominating bacteria should be gram-positive. This method also works on anaerobic bacteria. [Reference: H. C. Lichstein and M. H. Soule. *J. Bacteriol.* **47**, 221–30 (1944).]

2. Isolation of Temperature-Resistant or Spore-Forming Microorganisms

Materials

Water bath
Nutrient agar medium

Procedure

1. Incubate soil or water sample diluted in water in a water bath for enough time (30–60 minutes) to kill most microorganisms except

those that are temperature resistant (60°-100°C) or are spore formers (80°-100°C).
2. Streak heated sample on nutrient agar and incubate.

Observation

Only temperature-resistant cells or spores will survive heating and will multiply in the medium.

B. *Anaerobes*

3. Isolation of *Clostridium*

Materials

Nutrient medium	
Peptone	1.0 g
Lab-Lemco	1.0 g
Sodium acetate · 3 H_2O	0.5 g
Yeast extract	0.15 g
Soluble starch	0.1 g
Glucose	0.1 g
L-Cysteine	0.05 g
Agar	1.5 g
Water to	100 ml
pH (adjusted with NaOH)	7.1-7.2
Sodium sulfite	4.0%
Ferric citrate (heat to dissolve)	7.0%

Procedure

1. Filter-sterilize sodium sulfite and ferric citrate solutions (store sealed in refrigerator).
2. Melt nutrient medium.
3. Mix equal volumes of the sodium sulfite and ferric citrate solution and add to nutrient medium in the proportion of 1.0 ml of sodium sulfite-ferric citrate solution to 50 ml of nutrient medium.
4. Inoculate soil or water sample into melted, but cool, agar (45°-48°C) and pour into tube shown in diagram Fig. VII-1.
5. Incubate 20°-37°C for several days.

Observation

Anaerobic bacterial colonies appear as black iron sulfite areas (see Fig. VII-1). Agar butt may be pushed out of tube and colonies picked and streaked on nutrient medium and cultured anaerobically. The obligate anaerobic, gram-positive, spore-forming bacteria are *Clos-*

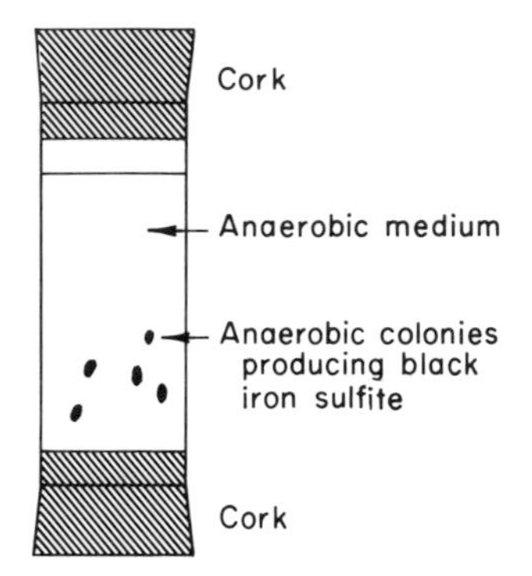

FIG. VII-1. Sealed anaerobic column.

tridium. [Reference: B. M. Gibbs and B. Freame. *J. Appl. Bacteriol.* **28**, 95–111 (1965).]

4. Isolation of Anaerobic Bacteria by the Plastic Film Technique

Materials

Nutrient agar medium

Tryptone	1.5 g
Yeast extract	1.0 g
Sodium thioglycollate	0.06 g
L-Cystine	0.01 g
Agar	1.5 g
Water to	96.5 ml
pH	7.1

Sterilize the following by filtration and add as indicated to above sterile nutrient agar:

$Na_2SO_3 \cdot 7\ H_2O$ (10,0%, aqueous)	0.5 ml
Polymyxin B (0.06%, aqueous)	1.0 ml
Sodium sulfadiazine (1.2%, aqueous)	1.0 ml
$FeSO_4 \cdot 7\ H_2O$ (5.0%, aqueous)	1.0 ml

Saran or equivalent thin plastic film (1 mm thick)

Cut plastic to fit over surface of agar plate. Pack in Petri dish with paper toweling separating plastic sheet. Sterilize with ethylene oxide at 1 atm in a dessicator for 24 hours at room temperature.

Procedure

1. Suspend sample in sterile water.
2. Dilute sample and add aliquots to Petri dish.
3. Prepare pour plate with nutrient agar medium supplemented with sulfite, Polymyxin B, sodium sulfadiazine, and ferrous sulfate and pour 12-15 ml into Petri dish.
4. Mix well; allow to harden.
5. Cover surface of plate with Saran and press to remove air pockets. Plastic does not permit oxygen to penetrate the agar readily.
6. Incubate for 48 hours at 20°-37°C (temperature varies with source of sample).
7. Sulfadiazine and Polymyxin tend to inhibit gram-negative bacteria and gram-positive cocci.

Observation

Anaerobic bacteria (clostridia) will produce black colonies (Fig. VII-2). Culture purity and authenticity of anaerobiosis may be confirmed by dilution and subculture both aerobically and anaerobically. [Reference: J. L. Shank. *J. Bacteriol.* **86**, 95-100 (1963).]

5. Enrichment and Isolation of Obligately Anaerobic Psychrophilic Bacteria

Materials

Trypticase soy broth and agar (or equivalent rich medium).
Anaerobic jars

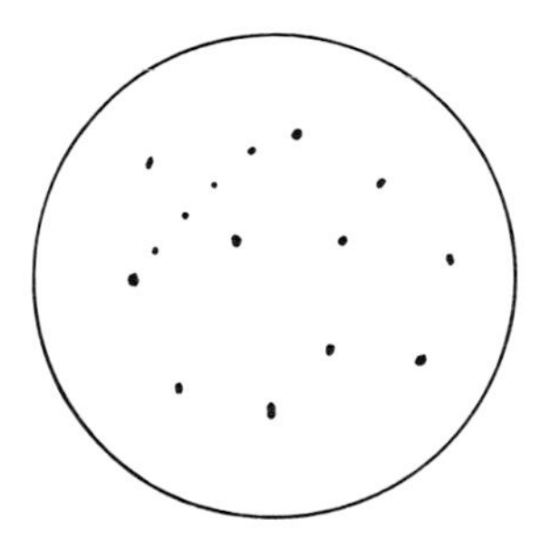

Fig. VII-2. Colonies appear black due to the deposition of black salts around the cells.

Procedure

1. Samples of ground meat, raw milk, mud, soil, or sewage are placed in duplicate in sterile 60-ml glass-stoppered bottles.
2. Fill bottles completely with freshly prepared, sterile Trypticase soy broth.
3. Heat one bottle to 80°C to kill vegetative cells; both bottles are then incubated at 0°C for 2 weeks.
4. Pure cultures may be isolated by streaking material from bottle showing bacterial growth onto the surface of Trypticase soy agar plates.
5. Incubate plates anaerobically at 0°C. Residual oxygen may be removed by adding 1-2 g of pyrogallic acid in 50-100 ml of 35% K_2CO_3 to the anaerobic dish before sealing.

Observation

Anaerobic psychrophilic bacteria will probably be gram-positive, spore-forming bacteria belonging to the genus *Clostridium*. [Reference: N. A. Sinclair, and J. L. Stokes, *J. Bacteriol.* **87**, 562-5 (1964).]

C. Marine Bacteria

6. Media for the Isolation of Marine Bacteria

Materials

ZoBell's 2216e medium	
Peptone	0.5 g
Yeast extract	0.1 g
$FePO_4$	0.01 g
Agar	1.5 g
Aged seawater to	100 ml
pH	7.5-7.8
Aerobic bacterial medium	
Casein hydrolyzate	0.5 g
K_2HPO_4	0.005 g
$FePO_4$	trace
Aged seawater	75 ml
Water to	100 ml
pH	7.6
Medium for sulfate-reducing bacteria	
Sodium lactate	0.35 g
NH_4Cl	0.1 g
K_2HPO_4	0.05 g

$MgSO_4 \cdot 7\,H_2O$	0.2 g
Ascorbic acid	0.01 g
Asparagine	0.005 g
Aged seawater	75 ml
Water to	100 ml
pH	7.6

Modified ZoBell's 2216e medium

Peptone	0.5 g
Yeast extract	0.1 g
Glucose	0.5 g
Ascorbic acid	0.01 g
$FePO_4$	0.001 g
Aged seawater	75.0 ml
Water to	100 ml
pH	7.6

Procedure

1. Agar (1.5 g/100 ml) may be added to solidify all of these media. The first two listed may be used for aerobic bacteria; the third for sulfate-reducing bacteria, and the last for anaerobic bacteria.
2. The liquid media are sterilized 10 ml per tube for aerobes and enough to fill container (tube or bottle) to the top for anaerobic or sulfate-reducing bacteria.
3. The bottles or tubes are inoculated with seawater sample and incubated in the dark at 18°–28°C (depending on source of sample) for several days or until there are signs of bacterial growth.

Observation

Bacterial growth is recognized by clouding of the liquid medium or colonies on agar Petri dishes. This may be confirmed by microscopic examination. [Reference: W. Gunkel, and C. H. Oppenheimer, *In* "Symposium on Marine Microbiology" (C. H. Oppenheimer, ed.), pp. 674–684. Thomas, Springfield, Illinois, 1963.]

7. MEDIA FOR THE ISOLATION OF EXTREMELY HALOPHILIC BACTERIA

Materials

Complex medium of Dundas

NaCl	25.0 g	Yeast extract	1.0 g
KCl	0.5 g	Water to	100 ml
$MgCl_2 \cdot 6\,H_2O$	0.5 g	pH	6.8
NH_4Cl	0.5 g		

Complex medium of Sehgal and Gibbons

Ingredient	Amount	Ingredient	Amount
Casamino acids	0.75 g	NaCl	25.0 g
Yeast extract	1.0 g	$FeCl_2$	2.3 mg
Sodium citrate	0.3 g	Water to	100 ml
KCl	0.2 g	pH	6.2
$MgSO_4 \cdot 7\,H_2O$	2.0 g		

Synthetic medium

Ingredient	Amount	Ingredient	Amount
DL-Alanine	43.0 mg	Cytidylic acid	10.0 mg
L-Arginine	40.0 mg	NaCl	25.0 g
DL-Aspartic acid	45.0 mg	$MgSO_4 \cdot 7\,H_2O$	2.0 g
L-Cystine	5.0 mg	KNO_3	10.0 mg
L-Glutamic acid	130.0 mg	K_2HPO_4	5.0 mg
Glycine	6.0 mg	KH_2PO_4	5.0 mg
DL-Histidine	30.0 mg	Sodium citrate	50.0 mg
DL-Isoleucine	44.0 mg	$FeCl_2$	0.23 mg
L-Leucine	80.0 mg	$CaCl_2 \cdot 7\,H_2O$	0.7 mg
L-Lysine	85.0 mg	$MnSO_4 \cdot 5\,H_2O$	0.03 mg
DL-Methionine	37.0 mg	$ZnSO_4 \cdot 7\,H_2O$	0.044 mg
DL-Phenylalanine	26.0 mg	$CuSO_4 \cdot 5\,H_2O$	5.0 μg
L-Proline	5.0 mg	Glycerol	0.1 g
DL-Serine	61.0 mg	Tween 40	50.0 mg
DL-Threonine	50.0 mg	Biotin	0.1 μg
L-Tyrosine	20.0 mg	Folic Acid	10.0 μg
DL-Tryptophan	5.0 mg	Vitamin B_{12}	0.02 μg
DL-Valine	100.0 mg	Water to	100 ml
Adenylic acid	10.0 mg	pH (adjusted	
Guanylic acid	10.0 mg	with KOH)	6.2
Uridylic acid	10.0 mg		

Procedure

1. Each of the abovementioned may be used with or without sterilization at 121°C for 20 minutes. Agar (1.5–2.0%) may be used for plates.
2. Samples of marine media, marine organisms, dried marine fish, solar salt, etc., are inoculated in 10 ml of medium in a 25–50-ml flask or tube and incubated at 37°C on a shaker or in stationary culture. Cultures are observed periodically.

Observation

Growth of bacteria in the above media is indicative of halophilic bacteria. These may be isolated as pure cultures on above media which has been hardened with agar. These bacteria do not stain with ordinary bacterial stains unless these are prepared with NaCl (25%).

[Reference: H. Onishi, M. E. McCance, and N. E. Gibbons. *Can. J. Microbiol.* **11**, 365-373 (1965).]

8. ISOLATION OF EXTREMELY HALOPHILIC BACTERIA*

Materials

Media (Agar, 1.8%, may be used to solidify)

a. Tryptone-yeast extract

Tryptone	0.5 g
Yeast extract (or autolyzate)	0.5 g
Solar salt†	25.0 g
$MgSO_4 \cdot 7\ H_2O$	2.0 g
$CaCl_2 \cdot 2\ H_2O$	0.5 g
Water to	100 ml

b. Fish broth peptone agar

Peptone	0.1 g
Glycerol	0.5 g
NaCl (or solar salt)	25.0 g
Agar	1.5 g
Fish broth* to	100 ml

c. Lochhead's skim milk agar

Skim milk (fresh)	50 ml
Agar	1.5 g
NaCl (pure or solar)	25.0 g
Water to	100 ml

Procedure

1. Enrichments may be made by placing a strip of salt fish on a wooden block floating in a 25% NaCl solution in a closed wide-mouth jar. Fish should not touch water.
2. Incubate 1-2 weeks at 37°C in dark.

Observation

Look for colored growth on fish. If present, streak colored material on any one of the nutrient agars in a Petri dish to isolate pure cultures. Cultures may be isolated from solar salt crystals placed on nutrient

**Warning:* halophilic bacteria are fragile and require concentrated salts in staining solutions or other experimental solutions to maintain their shape.

†Solar salt dissolved by boiling; debris removed by filtration.

*Fish broth is prepared by digesting 500 g of minced fresh cod filet with 500 ml of water. Filter digest through cheese cloth. Adjust volume to 1000 ml. Dilute broth 1:1 by volume before use.

agar plates. [Reference: K. Eimhjellen, 1965. *In* "Anreicherungskultur und Mutantenauslese" (H. G. Schlegel, ed.), pp. 126–37. Verlag, Stuttgart, 1965.]

9. Isolation of Marine Bacteria (after ZoBell and Feltham)

Materials

Nutrient medium A	
Bacto-peptone	0.5 g
Bacto-beef extract	0.2 g
KNO_3	0.05 g
Agar	1.5 g
Seawater to	100 ml
pH (after autoclaving)	7.8
Nutrient medium B	
Bacto-neopeptone	0.2 g
Bacto-tryptone	0.2 g
Bacto-peptone	0.2 g
Bacto-beef extract	0.2 g
Agar	1.5 g
Seawater to	100 ml
pH (after autoclaving)	7.8

Procedure

1. Autoclave either first or second medium. A precipitate may form which should be well mixed.
2. Add 1.0 ml of raw seawater to each Petri dish.
3. Allow agar medium to cool to approximately 45°C (or just before hardening) before pouring.
4. Pour and mix well and allow to harden.
5. Incubate in the dark at temperatures ranging from 0° to 37°C for 1–2 weeks.
6. Seal plates with Parafilm or plastic sandwich bag to prevent dessication.

Observation

Note difference in number of colonies as a function of incubation temperature. The second medium is better for prolonged cultivation and isolation of fastidious bacteria. Seawater samples stored at very low temperatures (4°C) yield a greater variety of marine bacteria. [Reference: C. E. ZoBell, and C. B. Feltham. *Bull. Scripps Inst. Oceanography, Univ. Calif. Tech. Ser.* **3**, 279–96 (1934).]

10. Isolation and Maintenance of Marine Heterotrophic Bacteria

Material

Medium

Trypticase	0.2 g
Soytone	0.2 g
Yeast extract	0.2 g
Agar	1.5 g
Marine mud extract	10.0 ml
Natural seawater	90.0 ml

Procedure

1. Mud extract is prepared by autoclaving 1 kg of wet marine mud in 1 liter of seawater for 20 minutes.
2. Filter.
3. Preserve filtrate under toluene or volatile preservative in refrigerator.
4. Prepare sterile agar plates containing nutrient medium.
5. Streak surface of plate with marine soil or water samples.

Observation

Look for bacteria growing on agar plate. Pure cultures may be isolated by repeated transfer of well-isolated colonies on fresh nutrient agar until pure. [Reference: P. Burkholder. *In* "Symposium on Marine Microbiology" (C. H. Oppenheimer, ed.), pp. 133–150. Thomas, Springfield, Illinois, 1963.]

11. Isolation of Halophilic Bacteria

Materials

Medium

NaCl	30 g
Proteose peptone*	1.0 g
Agar	1.5 g
Water to	100 ml

Procedure

1. Streak marine soil or water onto surface of above medium and incubate at 25–30°C for several weeks.
2. Seal plates with Parafilm or plastic to prevent evaporation.

*May be replaced by tryptone

Observation

Look for pigmented and nonpigmented colonies of rods, cocci, etc. [Reference: C.J.P. Spruit, and A. Pijper. *Antonie van Leeuwenhoek J. Microbiol. Serol.* **18**, 190-200 (1952).]

12. Isolation of Salt-Marsh Bacteria

Material

Basal nutrient medium	
Glucose	0.1 g
KNO_3	0.05 g
$FePO_4$	0.01 g
Artificial seawater to	100 ml
Artificial seawater	
NaCl	2.4 g
Na_2SO_4	0.4 g
$MgCl_2 \cdot 6\ H_2O$	1.1 g
$CaCl_2 \cdot 6\ H_2O$	0.2 g
KCl	0.07 g
$NaHCO_3$	0.02 g
KBr	0.01 g
$SrCl_2 \cdot 6\ H_2O$	0.004 g
H_3BO_3	0.003 g
$Na_2SiO_3 \cdot 9\ H_2O$	0.5 mg
NaF	0.3 mg
Demineralized water to	100 ml
Heterotrophic medium	
Vitamin-free casein hydrolyzate	0.4 g
Inositol	0.005 g
Thiamine	10 μg
Biotin	0.01 μg
Pyridoxine	20 μg
Pantothenic acid	10 μg
Nicotinic acid to	10 μg
Basal nutrient medium	100 ml
Medium S4	
Basal nutrient medium	75 ml
Soil extract (made with artificial seawater)	25 ml
Yeast extract	0.01 g
Agar (used to solidify above media)	1.5 g/100 ml

Procedure

1. Streak surface of any one of the above media solidified with agar in a Petri dish.
2. Incubate in the dark at 20–30°C. for 1 week.

Observation

Check plates for bacterial colonies. [Reference: T.R.G. Gray. *J. Gen. Microbiol.* **31**, 483-90 (1963).]

13. STERILITY TEST MEDIA FOR MARINE BACTERIA[a]

	STP	ST$_3$
Seawater	80 ml	70 ml
Water	15 ml	25 ml
Soil extract	5 ml	5 ml
$NaNO_3$	20.0 mg	5.0 mg
K_2HPO_4	1.0 mg	–
Na_2 glycerophosphate · 5 H_2O	–	1.0 mg
Hy-Case (Sheffield Chemical)	–	2.0 mg
Yeast extract (Difco)	–	1.0 mg
Yeast autolyzate (N.B.C.)	20.0 mg	–
Liver Oxoid L-25 (Oxo)	–	1.0 mg
Vitamin B_{12}	–	0.01 μg
Thiamine · HCl	0.02 mg	0.02 mg
Nicotinic acid	0.01 mg	0.01 mg
Putrescine · 2 HCl	4.0 μg	4.0 μg
Calcium pantothenate	0.01 mg	0.01 mg
Riboflavine	0.5 μg	0.5 μg
Pyridoxine · 2 HCl	4.0 μg	4.0 μg
Pyrodoxamine · 2 HCl	2.0 μg	2.0 μg
p-Aminobenzoic acid	1.0 μg	1.0 μg
Biotin	0.05 μg	0.05 μg
Choline H_2 citrate	0.05 mg	0.05 mg
Inositol	0.1 mg	0.1 mg
Thymine	0.08 mg	0.08 mg
Orotic acid	0.026 mg	0.026 mg
Folic acid	0.25 μg	0.25 μg
Folinic acid	0.02 μg	0.02 μg
Sucrose	100.0 mg	–
Na H glutamate	50.0 mg	–

[a]From M. Tatewaki, and L. Provasoli. *Botan. Marina* **6**, 193-203 (1964).

13. Sterility Test Media for Marine Bacteria[a] continued

	STP	ST_3
DL-Alanine	10.0 mg	2.0 mg
Trypticase (BBL)	20.0 mg	–
Glycine	10.0 mg	2.0 mg
Glycylglycine	–	40.0 mg
L-Asparagine	–	2.0 mg
Sodium acetate · 3 H_2O	–	4.0 mg
Glucose	–	5.0 mg
L-Glutamic acid	–	4.0 mg
pH	7.5-7.6	7.9

D. Bacteria Digesting Relatively Stable Plant Products

14. Enrichment and Isolation of Agar-Digesting Bacteria

Materials

Enrichment medium	
Agar	0.1 g
KNO_3 [or $(NH_4)_2SO_4$]	0.1 g
Seawater to	100 ml
Maintenance medium	
Peptone	0.5 g
$CaCO_3$	2.0 g
Agar	1.0-2.0 g
Aged seawater (or 3% NaCl in water) to	100 ml

Procedure

1. Inoculate a few milliliters of seawater or pieces of algal thalli into shallow layers of nonsterile or sterile enrichment medium in flasks.
2. Incubate 5-7 days at 22°C.
3. Streak material from enrichment medium onto surface of sterile maintenance medium in Petri dishes.
4. Incubate for 5-6 days at 22°C.

Observation

Look for areas of agar digestion around or under colonies produced by aerobic, gram-negative rods. Some are motile, either by creeping or with flagella. The liquefication of agar may be made visible by pouring I-KI solution over surface of agar; the unhydrolyzed agar stains reddish-violet while the hydrolyzed agar appears light straw color. This stain does not work on agar which has been solidified for several weeks. [Reference: R. Y. Stanier. *J. Bacteriol.* **42**, 527-559 (1941).]

15. Isolation of Marine and Soil Alginolytic Bacteria*

Materials

Nutrient medium (base layer)	
K_2HPO_4	0.03 g
Trypticase	0.25 g
Agar	1.5 g
Seawater to	100 ml
Nutrient medium (upper layer)	
Sodium alginate	2.5 g
NaCl	3.0 g
K_2HPO_4	0.03 g
Water to	100 ml

Procedure

1. A mixture of the dry NaCl and sodium alginate are added slowly to a solution of the K_2HPO_4 with vigorous stirring. This medium should have a pH of 7.0 after autoclaving 15 minutes at 121°C.
2. Pour 10–20 ml of sterile nutrient medium (base layer) into a Petri dish. Allow to harden.
3. Pour alginate nutrient medium (upper layer) over base layer to a depth of 0.5 cm and allow to harden at least 1 hour.
4. Remove water of syneresis of the alginate gel by drying at 37°C for 24–48 hours.
5. Streak marine mud or water over surface of alginate layer and incubate at 25°C for several days. *Do not break through to basal agar layer.*

Observation

Look for depression or pit in upper alginate layer around colony. Float alginate layer off agar base to confirm pit. [Reference: W. Yaphe. *Nature*, **196**, 1120–1 (1962).]

*For hydrolysis of alginate by soil bacteria substitute a soil extract agar medium for the base layer and dissolve sodium alginate (2.5 g) in aqueous K_2HPO_4 (0.03 g/100 ml) for the upper layer.

16. Enrichment and Isolation of Cellulolytic Microorganisms (Bacteria or Fungi)

Materials

250-ml flasks (cotton or screw-top)
Cellulose powder or finely shredded paper
Filter paper cut to fit Petri dish
Nutrient medium

Soil extract	0.05 ml
K_2HPO_4	0.05 g
$MgSO_4 \cdot 7\ H_2O$	0.05 g
NH_4Cl	0.05 g
Water (or seawater) to	100 ml
pH (approx.)	7.0

Agar 1.5 g/100 ml to nutrient medium to prepare plates.

Procedure

1. Prepare flasks with 100 ml of nutrient medium plus small pieces of filter paper.
2. Inoculate with 1.0 g of forest or garden soil, marine mud, or 5.0 ml of a water sample.
3. Minimize evaporation by covering top of flask with plastic.
4. Incubate for 1–3 weeks in dark at 25°–30°C.
5. Reinoculate new enrichment flask with fluid containing microorganisms from last flask. Repeat as necessary.
6. Streak bacterial fluid from enrichment flask onto surface of a sterile nutrient medium plate containing cellulose powder (1.0 g/100 ml medium) or sheet of filter paper overlayed by nutrient medium.

Observation

Look for colonies of bacteria or other microorganisms surrounded by a clear halo of digested paper (Fig. VII-3).

17. Enrichment and Isolation of Cellulolytic Bacteria and Fungi

Materials

Powdered cellulose
O-Phosphoric acid (85%)
Na_2CO_3 (1%, w/v)
Nutrient medium

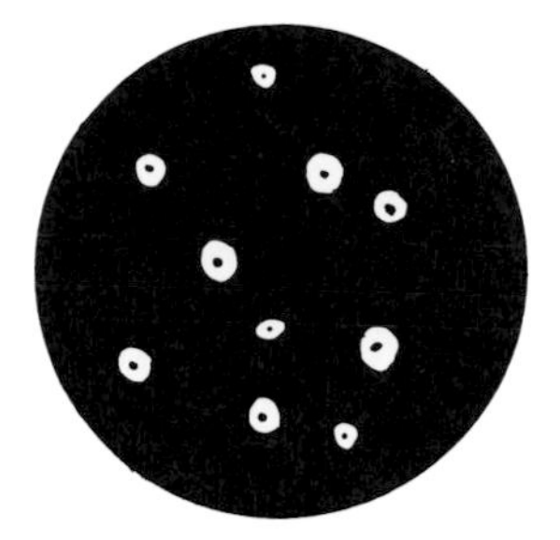

FIG. VII-3. Digestion of macromolecules by microbial enzymes seen as the development of clear or lighter areas around colonies.

$NH_4H_2PO_4$	0.2 g
KH_2PO_4	0.06 g
K_2HPO_4	0.04 g
$MgSO_4 \cdot 7\,H_2O$	0.089 g
Thiamine · HCl	10.0 μg
Yeast extract	0.05 g
Adenine	0.4 mg
Adenosine	0.8 mg
Treated cellulose	0.5 g
Agar	1.7 g
Water to	100 ml
250-ml flasks	

Procedure

1. Powdered cellulose is treated as follows: it is swollen in 85% *O*-phosphoric acid for 2 hours at 4°C, and then washed in the cold by repeated suspending and decanting with distilled water until neutral.
2. Prepare nutrient medium without agar and distribute 100 ml in 250-ml flasks.
3. Add 1.0 g soil or water sample.
4. Incubate for 7–35 days at room temperature.
5. Streak material from flasks on surface of solidified nutrient medium (with agar 1.7%); and incubate for 7–35 days at room temperature.

Observation

Look for colonies of fungi or bacteria surrounded by zone of clearing of cellulose (Fig. VII-3). [Reference: G. S. Rautela, and E. B. Cowling, *Appl. Microbiol.* **14**, 892–898 (1966).]

18. Enrichment and Isolation of Aerobic Cellulolytic Bacteria

Materials

Cellulose powder

Cellulose agar medium

Cellulose powder*	0.5 g
$NaNO_3$	0.1 g
$Na_2HPO_4 \cdot 7\ H_2O$	0.118 g
KH_2PO_4	0.09 g
$MgSO_4 \cdot 7\ H_2O$	0.05 g
KCl	0.05 g
Yeast extract	0.05 g
Casein hydrolyzate	0.05 g
Agar	1.0 g
Water to	100 ml
pH	6.8

Procedure

1. Enrich by mixing 100 g of soil with 5 g of powdered cellulose and incubate in a moist Petri dish (or other container) for 8 days at 28°C. Maintain humidity.
2. Isolate cellulolytic bacteria by streaking enrichment material on surface of agar-containing cellulose medium, making sure that all cellulose is well distributed.
3. Incubate for 8 days at 28°C.

Observation

Look for colonies with a tiny ring of cellulose digestion around them. Cellulolytic bacteria may be purified by growing them in a liquid medium containing Kanamycin sulfate (0.25–1 mg%) [Reference: J. Rivière, *Ann. Inst. Pasteur* **101**, 253–258 (1961).]

19. Isolation of Marine Chitinolytic Bacteria

Materials

Chitin medium

Precipitated chitin (or ground chitin)	0.1 g
Yeast extract	0.1 g

*Cellulose powder is treated for 12 hours with 1 *N* HCl in the cold and washed well to eliminate acid before it is used.

Agar	1.5 g
Aged seawater to	100 ml
pH	7.5–7.8

Procedure

1. Streak surface of sterile chitin agar medium with marine mud or water sample.
2. Incubate 4–40 days at 22°–25°C in the dark.

Observation

Look for colonies surrounded by clear zones of digested chitin (Fig. VII-3) [Reference: D. W. Lear, Jr., *In* "Symposium on Marine Microbiology" (C. H. Oppenheimer, ed.), pp. 594-610. Thomas, Springfield, Illinois, 1963.]

20. Enrichment for Chitin-Decomposing Bacteria

Materials

Decalcified chitin in strips
Chitin suspension (see below)
Mineral medium

K_2HPO_4	0.1 g
$MgSO_4 \cdot 7\,H_2O$	0.05 g
NaCl	0.05 g
$CaCl_2$	0.01 g
$FeCl_3$	0.005 g
Water to	100 ml
pH	7.0–7.2

Procedure

1. Decalcify lobster exoskeleton in dilute HCl.
2. Boil chitin for 12 hours in 10% KOH and wash.
3. Treat chitin with dilute $KMnO_4$ for 20 minutes at 60°C; wash, and suspend in cold concentrated sodium bisulfite until decolorized.
4. Wash well to remove bisulfite and dry at 80°C. This should result in leathery strips of chitin.
5. Dissolve a strip of chitin in cold 50% H_2SO_4 and reprecipitate by 10- to 20-fold dilution in distilled water.
6. Wash precipitate free of acid by repeated water washings and centrifugation. The resulting product must be kept as a sterile aqueous suspension.

7. Enrichment cultures are set up by adding soil or mud to mineral medium in a shallow layer in a Petri dish or bowl containing thin strips of chitin and then incubating at 25°–30°C for 7–14 days.
8. Streaks from enrichment cultures are made on agar plates containing mineral medium and finely divided chitin and hardened with 2% agar.

Observation

Chitinolytic bacteria will appear as follows: (1) In a few days pseudomonads producing small white or pale yellow colonies surrounded by cleared zones in the agar may appear. Chitinolytic *Streptomyces* spp. may also appear. (2) After 5 or 6 days chitin-decomposing myxobacteria appear as thin translucent, pale yellow, almost colorless swarms which spread and cover large areas of the plate. These myxobacteria do not seem to be chitinlytic in young cultures. They are best seen in young cultures by reflected light when their matt surface shows up in contrast to the surrounding agar. Pure cultures of myxobacteria may be isolated by streaking these bacteria on agar containing 0.5% tryptone or on mineral agar medium supplemented with a chitin suspension. Cells are long (8–12 μ), flexible, and show creeping movement in young cultures. Cells swarm on solid media containing 0.25% or less tryptone and form colonies in 1.0% or more tryptone. [Reference: R. Y. Stanier, *J. Bacteriol.* **53**, 297–315 (1947).]

21. Enrichment and Isolation of Chitinolytic Microorganisms (Bacteria or Fungi)

Materials

250-ml flasks (cotton or screw-top)
Chitin from exoskeleton of shrimp, lobster, crab, crayfish, etc.
Mineral medium

K_2HPO_4	0.1 g
$MgSO_4 \cdot 7\,H_2O$	0.05 g
NaCl	0.05 g
$CaCl_2$	0.01 g
Fe [as $Fe(NH_4)_2(SO_4)_2 \cdot 6\,H_2O$]	0.5 mg
NH_4Cl	0.1 g
Water to	100 ml
pH	7.0

Agar mineral medium (add 1.5 g agar per 100 ml medium to solidify).

HCl	1%
KOH	2%
Ethanol	95%
H_2SO_4	50%

Büchner funnel

Procedure

1. Chitin is prepared by washing crustacean exoskeleton in warm water. Decalcify chitin in 1% (v/v) HCl. Change HCl each day for 4 days. Wash chitin to remove HCl. Place chitin in 2% (w/v) KOH for 10 days. Bring chitin plus KOH to a boil and allow to cool. Repeat KOH and boiling and cooling 4 times during 10-day period. Wash chitin free of KOH. Cut into strips and extract 4 times with boiling ethanol. Dry strips.
2. Several strips of purified chitin are added to 50 ml of mineral medium in a 250-ml flask. Flask is inoculated with marine mud and incubated for 1-3 weeks in the dark at 25°-30°C. After 2-3 weeks fresh chitin-containing flasks (as above) are inoculated from previous flask to enrich for chitinolytic microorganisms. This may be repeated several times.
3. Chitin agar plates are made by dissolving chitin strips in 50% (v/v) H_2SO_4 and precipitating the chitin by a 15-fold dilution with water. The chitin precipitate is washed free of acid on a Büchner funnel. *Do not allow chitin precipitate to dry.* The wet chitin is added to the mineral medium plus agar and sterile plates are prepared.

Observation

Chitinolytic microorganisms will appear in the enrichment flasks by producing halos of decomposition in the chitin strips. Material from these enriched flasks is streaked on the surface of the chitin-agar-mineral medium plates and these are incubated 1-3 weeks in the dark at 25°-30°C or until zones of clearing appear around colonies. These are the chitinolytic microorganisms. [Reference: L. L. Campbell, Jr. and O. B. Williams, *J. Gen. Microbiol.* **5**, 894-905 (1951).]

22. Enrichment and Isolation of Lignanolytic Bacteria

Materials

K medium

$NaNO_3$	0.25 g
KH_2PO_4	0.1 g
$CaCl_2$	0.001 g
$MgSO_4 \cdot 7\ H_2O$	0.03 g
NaCl	0.01 g
$FeCl_3$	0.001 g
Thiamine	0.25 mg
Calcium pantothenate	0.25 mg
Biotin	0.5 μg
Vitamin B_{12}	1.0 μg
Pyridoxine	0.25 mg
Nicotinic acid	0.25 mg
p-Aminobenzoic acid	0.25 mg
Folic acid	0.25 mg
Water to	100 ml
pH	6.5

KYD Nutrient

Yeast extract	0.1 g
Agar	2.0 g
K medium to	100 ml

α-Conidendrin (from Crown & Zellerbach Corp., Camas, Washington). White crystals dissolved in acetone and reprecipitated from water.

Procedure

1. Soil samples (1.0 g) rich in wood residue (from decaying wood or sawdust) are inoculated into K medium containing 0.5% α-conidendrin (100 ml) in a 250-ml flask, for enrichment.
2. The enrichment procedure is repeated 3 times by transferring inocula from one enrichment to a fresh enrichment flask.
3. Decomposition of α-conidendrin is indicated by a darkening of the medium. Material from the last enrichment may be streaked on enrichment K medium solidified with agar (2%).
4. Incubate for several days at 20°–30°C.

Observation

Colonies surrounded by areas of darkening of the agar medium contain bacteria able to use α-conidendrin as sole carbon and energy source. Purify by repeated isolation on K medium plates with α-conidendrin as sole source of carbon and energy. Cultures may be maintained on KYD medium. [Reference: V. Sundman, *J. Gen. Microbiol.* **36**, 171–183 (1964).]

E. Bacteria Utilizing Hydrocarbon and Aromatic Molecules

23. Enrichment and Isolation of Methane-Utilizing Bacteria

Materials

Medium	
$NaNO_3$	0.2 g
$MgSO_4 \cdot 7\,H_2O$	0.02 g
$FeSO_4 \cdot 7\,H_2O$	0.1 mg
Na_2HPO_4	0.02 g
$NaH_2PO_4 \cdot H_2O$	9.0 mg
Cu (as $CuSO_4 \cdot 5\,H_2O$)	0.5 μg
B (as H_3BO_3)	1.0 μg
Mn (as $MnSO_4 \cdot 5\,H_2O$)	1.0 μg
Zn (as $ZnSO_4 \cdot 7\,H_2O$)	7.0 μg
Mo (as MoO_3)	1.0 μg
KCl	4.0 mg
$CaCl_2$	1.5 mg
Water to	100 ml
pH	7.0
Methane	

Procedure

1. Sterilize 50 ml of above medium in 250-ml flask.
2. Inoculate with 1.0 g of mud, soil, or sewage sample.
3. Incubate in an atmosphere of methane and air (50:50%) for 6–10 days at 30°–37°C.

Observation

Look for pink to brown pellicle on surface of medium. Pink pellicle appears at 30°C; brown pellicle at 37°C. Bacterial cells are gram-negative motile rods. Pure cultures may be isolated on Petri dishes containing mineral medium hardened with agar (2.0 g/100 ml). [Reference: E. R. Leadbetter, and J. W. Foster, *Arch. Mikrobiol.* **30**, 91–118 (1958).]

24. Enrichment and Isolation of Bacteria Utilizing Hydrocarbons

Materials

Mineral medium for freshwater*

NH_4Cl	0.05 g
$NaH_2PO_4 \cdot H_2O$	0.05 g
KH_2PO_4	0.05 g
$MgSO_4 \cdot 7\ H_2O$	0.05 g
NaCl	0.4 g
Water to	100 ml
pH	7.0

Procedure

1. For enrichment, sterilize mineral medium in a shallow layer in a large flask.
2. Sterilize phosphate salts separately and combine them aseptically with the medium to avoid precipitation.
3. Add hydrocarbon to make 1% and 1 g soil, or 10 ml water sample preferably from soil or water exposed to hydrocarbons.
4. Shake or aerate at 32°C for 1 week.
5. Bacteria utilizing hydrocarbons as their sole source of carbon may be isolated by streaking material from the enrichment culture onto mineral medium hardened with 2% agar in a Petri dish.
6. The volatile hydrocarbon may be added to the inverted cover of the Petri dish and the inverted dish incubated for 1 week or more at 32°C.

Observations

Bacteria able to grow on mineral medium in the presence of hydrocarbon, but not in its absence, are hydrocarbon-utilizing bacteria. [Reference: M. Konovaltschikoff-Mazoyer, and J. C. Senez, *Ann. Inst. Pasteur* **91**, 60–67 (1956).]

25. Enrichment and Isolation of Pseudomonads Oxidizing Naphthalene

Materials

250-ml flask
Medium (add agar 1.5 g/100 ml to harden)

*For marine bacteria add NaCl 2.0% and 10 ml aged seawater.

NH_4NO_3	0.4 g
K_2HPO_4	0.1 g
KH_2PO_4	0.05 g
$MgSO_4 \cdot 7\,H_2O$	0.05 g
$MnCl_2 \cdot 4\,H_2O$	0.02 g
$CaCl_2 \cdot 2\,H_2O$	0.005 g
$FeCl_2 \cdot 4\,H_2O$	0.005 g
$CaCO_3$	0.2 g
Yeast extract	0.01 g
Naphthalene	1.0 g
Water to	100 ml

Procedure

1. Add 1.0 g of soil or mud to shallow layer of above liquid medium or aerate larger volume in a 250-ml flask.
2. Incubate for 2–3 days at 25°–30°C or until brownish-black to reddish-orange pigment develops.
3. Isolate pigment-producing colonies by streaking liquid from enrichment flask on surface of above agar medium.

Observation

Microorganisms oxidizing naphthalene produce colonies which contain or diffuse brown to reddish-orange pigments (napthoquinones). [Reference: J. F. Murray, and R. W. Stone, *Can. J. Microbiol.* **1**, 579–588 (1955).]

26. Enrichment and Isolation of Microorganisms Oxidizing Toluene (or other Aromatic Compounds)

Materials

Soil
Toluene (or other aromatic compounds)
250-ml flask
Medium A-1 (agar, 2.0 g/100 ml may be added to solidify medium)

$(NH_4)_2SO_4$	0.12 g
$CaCl_2 \cdot 2\,H_2O$	0.01 g
$MgSO_4 \cdot 7\,H_2O$	0.01 g
Ferric citrate	0.2 mg
Water to	100 ml

Medium A-2	
K_2HPO_4	0.1 g
KH_2PO_4	0.05 g
Water to	100 ml
Medium B (agar (2.0 g/100 ml) may be added to solidify medium).	
K_2HPO_4	0.08 g
KH_2PO_4	0.02 g
$MgSO_4 \cdot 7\,H_2O$	0.05 g
$CaSO_4 \cdot 2\,H_2O$	5.0 mg
$FeSO_4 \cdot 7\,H_2O$	1.0 mg
$(NH_4)_2SO_4$	0.1 g
Water to	100 ml

Procedure

1. Sterilize medium A-1 and A-2 separately and mix together aseptically (A-1:A-2, 5:1, v/v, respectively).
2. Expose fresh soil to toluene (or other aromatic compound) vapors for several days.
3. Inoculate 1.0 g of toluene-exposed soil in a 250-ml flask containing 50-100 ml of either medium A-1 + A-2 (5:1, v/v) or medium B.
4. The flask is incubated in a desiccator (or other closed container) containing a beaker of water saturated with toluene for 1–3 weeks at 25°–30°C, or until growth appears in the flask.
5. Pure cultures are isolated by streaking from enrichment flasks onto agar plates prepared with medium A-1 + A-2 (5:1, v/v) or medium B.
6. Incubate plates at 25°–30°C in a sealed container containing a beaker of water saturated with toluene until colonies appear on agar.

Observation

Look for colonies of bacteria capable of growing in mineral media with toluene or other aromatic compounds as the sole source of carbon. [Reference: D. Claus, and N. Walker, *J. Gen. Microbiol.* **36**, 107–22 (1964).]

27. Enrichment and Isolation of Benzene-Oxidizing Bacteria

Materials

Isolation medium

NH_4NO_3	0.4 g
K_2HPO_4	0.1 g
KH_2PO_4	0.05 g
$MgSO_4 \cdot 7\,H_2O$	0.05 g
$MnCl_2 \cdot 4\,H_2O$	0.02 g
$CaCl_2 \cdot 2\,H_2O$	0.05 g
$CaCO_3$	0.2 g
Water to	100 ml
pH	7.0
Growth medium	
$NaNH_4HPO_4$	0.15 g
KH_2PO_4	0.1 g
$MgSO_4 \cdot 7\,H_2O$	0.02 g
$FeCl_2$	trace
Water to	100 ml
pH	6.8–7.0
Nutrient agar	

Procedure

1. Place soil (0.1 g) in a 500-ml Erlenmeyer flask containing 50 ml isolation medium plus 0.1 g benzene.
2. Shake in a reciprocal shaker at 30°C for several days.
3. Streak bacteria from isolation medium on nutrient agar plates.
4. Incubate 24 hours and inoculate colonies into tubes containing 5 ml of growth medium plus 0.005 g of benzene added just prior to inoculation.

Observation

Bacteria growing in mineral growth medium containing only benzene as carbon source are likely to be benzene-oxidizers. [Reference: E. K. Marr, and R. W. Stone, *J. Bacteriol.* **81**, 425-530 (1961).]

28. Enrichment and Isolation of Phenol-Utilizing Bacteria

Materials

Nutrient medium	
K_2HPO_4	0.1 g
$MgSO_4 \cdot 7\,H_2O$	0.02 g
NaCl	0.01 g
$CaCl_2$	0.01 g
$FeCl_3$	0.002 g

$(NH_4)_2SO_4$	0.1 g
Phenol	0.007 *M*
Water to	100 ml
pH	7.2–7.5

Procedure

1. Inoculate 1 g of soil into 250-ml Erlenmeyer flask containing 50 ml of nutrient medium.
2. Incubate for 3 days at 30°C.
3. Transfer 0.1 ml of liquid to a second sterile flask and incubate for 3 days at 30°C. Repeat this procedure 3 times.
4. At the end of this procedure, streak some of the enriched material from the last flask on the surface of an agar plate containing the nutrient medium plus agar (2%).

Observation

A number of different bacteria: micrococci, vibrios, achromobacters, etc. may appear on agar. Test these in liquid nutrient medium to see if they can grow with phenol as sole carbon source. Other aromatic compounds may be used in place of phenol. [Reference: N. Kramer, and R. N. Doetsch. *Arch. Biochem. Biophys.* **26**, 401–405 (1950).]

F. Bacteria Digesting Biopolymers or Producing Specific Enzymes

29. Isolation of Casein-Digesting Bacteria

Materials

Nutrient agar
Skim milk

Procedure

1. Prepare nutrient agar plates containing 1% (w/v) skim milk.
2. Streak surface with soil, water, or bacteria.

Observation

Look for colonies of microorganisms surrounded by a clear halo of digested casein (Fig. VII-3).

30. Isolation of Starch-Digesting (Amylase) Bacteria and Fungi

Materials

Nutrient agar medium containing 0.5% soluble starch

Gram's iodine

Iodine	1.0 g
KI	2.0 g
Water to	100 ml

Procedure

1. Streak surface of any nutrient agar medium containing soluble starch with soil or water sample.
2. Incubate until colonies are clearly visible.

Observation

Flood plate with Gram's iodine and pour off excess. Amylase-producing colonies will have a zone of brown color surrounding them, while prevailing agar appears blue to black. Color fades. Clones may be isolated by quickly streaking material from positive colonies on to fresh nutrient agar.

31. Isolation of Carbohydrate-Catabolizing Enzyme Producers

Materials

Carbohydrate(s)–10% aqueous solution

Nutrient agar medium

Tryptone	0.05 g
Yeast extract	0.025 g
L-Arginine	5.0 mg
Guanine	5.0 mg
Agar	1.7 g
Water (buffered with 0.12 *M* Tris to pH 7.5) to	100 ml

Chloramphenicol 0.5% aqueous solution

Triphenyltetrazolium chloride 4.0% in 1.0 *M* phosphate buffer at pH 7.0

Atomizers (2)

Procedure

1. Sterilize agar for 10 minutes at 15 lbs pressure. (Too much heating produces reducing compounds which interfere with reduction of dye.)
2. Streak surface of agar plate with soil or water sample to obtain isolated colonies.
3. Incubate plates until well-formed colonies appear. Avoid long incubation.
4. Spray plate culture with carbohydrate solution (10%).

5. Incubate for 30 minutes at incubation temperature of microorganisms.
6. Spray plate with 4% triphenyltetrazolium chloride. Colonies producing enzymes dehydrogenating the carbohydrate substrate will accumulate reducing substances which reduce the triphenyltetrazolium to a red color.
7. The red colonies may be streaked on fresh medium and clones isolated.
8. The induction of adaptive enzymes may be avoided by spraying the plate with chloramphenicol (0.5%) immediately after the substrate to prevent protein synthesis. [Reference: E.C.C. Lin, S. A. Lerner, and S. E. Jorgensen. *Biochim. Biophys. Acta* **60**, 422–424 (1962).]

32. Isolation of Lipolytic Microorganisms No. 1

Materials

Nutrient medium	
Tributyrin	1.0 g
Agar	1.5 g
Nutrient broth to	100 ml

Procedure

1. Blend tributyrin into nutrient broth with hand mechanical emulsifier to form a stable emulsion. Emulsifiers such as several milliliters of Tween 80 may be necessary.
2. Add agar.
3. Sterilize and pour agar medium into Petri dishes. Mix well before agar hardens.
4. Streak agar surface with soil or water sample.
5. Incubate for 3–7 days in the dark at 20°-37°C.

Observation

Look for colonies surrounded by clear area where tributyrin has been hydrolyzed (Fig. VII-3). [Reference: M. E. Rhodes. *J. Gen. Microbiol.* **21**, 221-63 (1959).]

33. Isolation of Lipolytic Microorganisms No. 2

Materials

Medium	
$NH_4H_2PO_4$	0.1 g

KCl	0.02 g
$MgSO_4 \cdot 7\ H_2O$	0.02 g
Yeast extract	0.3 g
Agar	2.0 g
Olive oil (or other fat or oil)*	5.0 g
Water to	100 ml
pH (adjust with NaOH)	7.8

Control medium (medium above minus olive oil)

Procedure

1. Autoclave above medium and pour into Petri dishes.
2. Mix well before allowing agar to harden.
3. Streak surface of olive oil agar medium with soil or water sample.
4. Incubate for 3–7 days in the dark at 20°–37°C.

Observation

Look for colonies surrounded by blue salt formed by reaction of free fatty acid and weak base of the dye. Streak positive colonies on both olive oil agar and control agar. Only lipolytic microorganisms give ring of blue salt around colonies in olive oil agar. [Reference: M. E. Rhodes. *J. Gen. Microbiol.* **21**, 221–263 (1959).]

34. Isolation of Esterase-Active Microorganisms

Materials

Filter paper discs soaked with indoxyl acetate (in ethanol) 0.05 mg/disc (can be alternatively soaked in indoxyl butyrate or β-naphthyl acetate)

Bacillus cereus is positive for enzyme.

Procedure

1. Streak organism over surface of nutrient agar plate.
2. Incubate until there is good surface growth.
3. Place disc soaked with indoxyl acetate on surface.

Observation

If cells produce an esterase the indoxyl acetate will be hydrolyzed to a product which is rapidly oxidized in the air to indican. Indican has a deep blue color. [Reference: P. H. Clarke, and K. J. Steel. *In*

*Nile blue (basic dye) 1 part by weight in 15,000 in final medium. Dissolve, however, in olive oil.

"Identification Methods for Microbiologists" (B. M. Gibbs and F. A. Skinner, eds.), pp. 111–115. Academic Press, New York, 1966.]

35. Isolation of β-Galactosidase-Active Microorganisms

Materials

Lactose, 15-mg, imbedded-filter paper disc
o-Nitrophenyl-β-D-galactoside, 0.1 mg/filter paper disc
Escherichia coli is positive for enzyme.

Procedure

1. Streak organism over surface of nutrient agar plate.
2. Incubate until there is good surface growth.
3. Place inducer disc, lactose, on surface and incubate for 2 hours at 37°C to induce β-galactosidase synthesis.
4. Place the substrate disc, *o*-nitrophenyl-β-D-galactoside, so it overlaps lactose disc.
5. Enzyme hydrolyzes the substrate producing *o*-nitrophenol which is yellow in color.
6. The reaction develops in ca. 15 minutes. [Reference: P. H. Clarke, and K. J. Steel. *In* "Identification Methods for Microbiologists" (B. M. Gibbs and F. A. Skinner, eds.), pp. 111–115. Academic Press, New York, 1966.]

36. Isolation of Phosphatase-Active Microorganisms

Materials

Prepare sterile solution of 1% phenolphthalein diphosphate (pentasodium salt) by filtration.
Nutrient agar medium + 1.0 ml of 1% phenolphthalein diphosphate

Procedure

1. Streak surface of agar medium with organism.
2. Incubate until there is good surface growth.

Observation

Invert agar plate over a Petri dish lid containing ammonia (sp. gr. 0.88). Colonies of phosphatase-producing organisms turn deep pink on exposure to ammonia vapors. [Reference: A. C. Baird-Parker. *In* "Identification Methods for Microbiologists" (B. M. Gibbs and F. A. Skinner, eds.), pp. 59–64. Academic Press, New York, 1966.]

37. Isolation of Amino Acid Oxidase-Active Microorganisms

Materials

L-Phenylalanine (filter paper disc soaked in 1% aqueous solution). (Other L-amino acids may be used.)
Ferric chloride, 10% in dilute HCl
Proteus sp. is positive for enzyme.

Procedure

1. Streak organism over surface of nutrient agar plate.
2. Incubate until there is good surface growth.
3. Place disc soaked with L-phenylalanine on surface for 2 hours at 37°C.

Observation

Add 1 drop of ferric chloride solution to disc. A deep green-blue color develops after 1 minute; color will turn darker with time. Color is a positive test for α-keto acids produced by amino acid oxidase. Phenylpyruvic acid is produced from L-phenylalanine. [Reference: P. H. Clarke, and K. J. Steel. *In* "Identification Methods for Microbiologists" (B. M. Gibbs and F. A. Skinner, eds.), pp. 111-115. Academic Press, New York, 1966.]

38. Isolation of Tryptophanase-Active Microorganisms

Materials

Filter paper disc soaked in 1% aqueous solution of tryptophan
Kovacs reagent (5% *p*-dimethylaminobenzaldehyde in 75 ml isoamyl alcohol + 25 ml concentrated HCl)
Escherichia coli is positive for enzyme

Procedure

1. Streak organism over surface of nutrient agar plate.
2. Incubate until there is good surface growth.
3. Place disc soaked with tryptophan on surface.
4. Incubate for 2 hours at 37°C.

Observation

Add 1-2 drops of Kovacs reagent to disc. A positive reaction gives a red color (which fades in *ca.* 15 minutes) due to the presence of indole. Tryptophanase converts tryptophan to indole. [Reference: P. H.

Clarke, and K. J. Steel. *In* "Identification Methods for Microbiologists" (B. M. Gibbs and F. A. Skinner, eds.), pp. 111-115. Academic Press, New York, 1966.]

39. Demonstration of Lysine Decarboxylase Activity in Bacteria

Materials

Nutrient medium

Casein hydrolyzate (enzymatic)	1.5 g
K_2HPO_4	0.2 g
Glucose	0.1 g
Water to	100 ml

Assorted bacteria
NaOH, 4 *N*
Chloroform
Ninhydrin (1,2,3-triketohydrindene), 0.1% in chloroform

Procedure

1. Inoculate bacteria into 5.0 ml of nutrient medium per tube.
2. Incubate until there is good growth.
3. Add 1 ml 4 *N* NaOH; mix.
4. Add 2 ml chloroform and shake vigorously.
5. Cadaverine, the decarboxylation product of lysine, appears in the chloroform (top) layer.
6. Centrifuge, if necessary, to break emulsion.
7. Remove 0.5 ml of clear chloroform layer to clean tube (13 × 100 mm).
8. Add 0.5 ml of 0.1% ninhydrin in chloroform and keep reaction mixture at room temperature.

Observation

After 4 minutes a positive reaction for cadaverine is indicated by a deep purple color. Incubation at higher temperatures or for longer time may give false positives. [Reference: P. R. Carlquist. *J. Bacteriol.* **71**, 339-341 (1956).]

40. Demonstration of Oxidase in Bacteria

Whatman No. 1 filter paper, 6-cm square
Tetramethyl-*p*-phenylenediamine · 2 HCl (1.0%, aqueous)

Procedure

1. Soak 6-cm square of filter paper in Petri dish with 2–3 drops of 1% tetramethyl-*p*-phenylenediamine · 2 HCl.
2. Wipe test colony on surface of paper with glass rod or inoculation needle.

Observation

Bacterial material turns dark purple in 5–10 seconds if oxidase present. [Reference: N. Kovacs. *Nature.* **178**, 703 (1956).]

41. Demonstration of Cytochrome Oxidase Activity

Materials

Bacteria
Nutrient broth
p-Aminodimethylaniline oxalate (1%, aqueous)
α-Naphthol (1%, in ethanol)

Procedure

1. Inoculate bacteria into nutrient broth and incubate 12–18 hours.
2. Place 2–5 ml of the broth culture in a clean tube.
3. Add 0.3 ml of *p*-aminodimethylaniline oxalate (1.0%) and 0.2 ml of α-naphthol (1%) and shake vigorously.

Observation

Appearance of a blue color indicates the presence of cytochrome oxidase. [Reference: W. L. Gaby, and E. Free. *J. Bacteriol.* **76**, 442–44 (1958).]

42. Benzidine Test for the Detection of Cytochrome-Containing Respiratory Systems in Microorganisms.

Materials

Microbial cultures
Benzidine · 2 HCl (1 g) is partially dissolved in 20 ml glacial acetic acid. Water (30 ml) is added and the solution is heated gently and cooled. Add 50 ml of ethyl alcohol (95%). Stable in refrigerator for 1 month.
Hydrogen peroxide, 30% reagent grade

Procedure

1. Place microbial cell material at bottom of tube.
2. Add 0.5 ml benzidine reagent and 0.5 ml of freshly prepared solution of hydrogen peroxide (5%).
3. Agar plates may be flooded with reagents as above.

Observation

A positive test is indicated by cells or colonies becoming blue to blue-green in color. [Reference: R. H. Deibel, and J. B. Evans. *J. Bacteriol* 79, 356–360 (1960).]

43. Isolation of Penicillinase-Producing Microorganisms

Materials

Dye, *N*-phenyl-1-naphthylamine-azo-*o*-carboxybenzene.*
 Dissolve 1.25 g of dye in 470 ml of dimethyl formamide. Add 30 ml 0.1 *N* NaOH. Stable indefinitely. *Dimethyl formamide is toxic.*
Penicillin G (2% aqueous). Unstable. Prepare daily and refrigerate.
Nutrient agar medium, pH 7.0

Procedure

1. Streak surface of nutrient agar in Petri dishes with soil, water, or microbial sample.
2. Incubate until colonies are conspicuous.
3. Dry plates with covers off for 2 hours at 37°C.
4. Flood each plate with 3 ml of dye in hood (*dimethyl formamide is toxic*).
5. Allow plates to remain in contact with the dye for 2 hours with the covers on.
6. Pour off dye and dry plates in vertical position for ca. 10 minutes.
7. Flood plates with 2–3 ml of 2% penicillin solution.
8. Color will develop in 1-2 minutes.

Observation

Look for purple colonies. Dye is an acid-base indicator which turns from yellow to purple as the pH drops from 7 to 5. The acid form is purple and insoluble; it precipitates in and on the positive colonies. Penicillinase hydrolyzes penicillin releasing an acid group which

*Available from Gallard-Schlesinger Chem. Corp., Carle Place, New York 11514.

lowers the pH and precipitates the dye. [Reference: R. P. Novick, and M. H. Richmond. *J. Bacteriol.* **90**, 467–480 (1965).]

G. Pigment-Producing Bacteria

44. Isolation of Fluorescent Pigment-Producing Bacteria

Materials

Nutrient medium

Yeast extract	0.3 g
Peptone	1.0 g
NaCl	0.5 g
Agar	2.0 g
Water to	100 ml
pH	7.2–7.4

Procedure

1. Streak soil or water samples on the surface of above nutrient medium in Petri dishes.
2. Incubate for 3–7 days in the dark at 25°C.

Observation

Examine colonies for characteristic green-yellow water-soluble fluorescin. Streak pigmented colonies on fresh agar medium to purify cultures. [Reference: M. E. Rhodes. *J. Gen. Microbiol.* **21**, 221–63 (1959).]

45. Enrichment and Isolation of Bacteria which Oxidize Indole to a Blue Pigment

Materials

Nutrient agar Petri dish containing 10.0 mg% indole or tryptophan 10.0 mg%

Tryptophan medium

Yeast extract	0.3 g
DL-Tryptophan	0.5 g
Na_2HPO_4	0.1 g
NaCl	0.5 g
Agar	1.2 g
Water to	100 ml

Procedure

1. Streak surface of nutrient agar or tryptophan agar medium in a Petri dish with soil or soil in water suspension.
2. Incubate for 48–96 hours in the dark at 25–30°C.

Observation

Look for colonies producing a blue pigment, indigotin, by oxidizing indole. These are likely to be *Pseudomonas indoloxydans*. This may be confirmed by isolating the organism and growing it on media free (or low) in indole or tryptophan. If pigment is produced only in the presence of indole or tryptophan, organism is an indole oxidizer. Gelatin is a protein with little tryptophan and no indole. [Reference: M. Polster. *Experientia* **22**, 360 (1966).]

46. Enrichment and Isolation of Chromobacteria

Materials

Beijerinck's fibrin medium	
Dried blood fibrin (egg white may be substituted)	1–2 g
KCl	0.02 g
Agar	3.0 g
Water to	100 ml
Rice grains, sterilized	
Rice agar	
Boiled and mashed rice	1.0 g
Agar	1.5 g
L-Tryptophan	2.5 mg
Water to	100 ml
Nutrient agar (BBL, Difco, etc.)	

Procedure

1. Enrichment may be accomplished by streaking soil or water samples on Beijerinck's medium or placing soil or water sample in Petri dish containing about 50 grains of sterile rice in 10–25 ml sterile water.
2. Incubate at 20°–25°C for 5–10 days.

Observation

Violet patches on the rice or fibrin agar may be chromobacteria. These are isolated by streaking violet material on the surface of rice agar (preferred medium) or nutrient agar. Look for violet-pigmented colonies. [Reference: Personal communication from Dr. P. H. A. Sneath, 1968, University of Leicester, Leicester, Great Britain.]

H. Iron, Sulfur, and Nitrogen Bacteria

47. ENRICHMENT FOR IRON BACTERIUM, *Sphaerotilus* NO. 1

Materials

Hay (alfalfa or other plant leaves)
$FeSO_4$
NaOH
$MnCO_3$
100-ml graduated or long glass cylinder

Procedure

1. Boil 10 gram alfalfa hay (or other plant leaves) for ½ hour in large volume of water. Discard liquid and save extracted plant material. Extracted leaves may be dried and saved.
2. Prepare ferric hydroxide $Fe(OH)_3$, just before use, by adding concentrated NaOH solution dropwise to a saturated $FeSO_4$ solution until no more $Fe(OH)_3$ is formed. Wash precipitate [$Fe(OH)_3$] on filter paper on Büchner funnel with distilled water to remove NaOH (or until wash water is no longer alkaline).
3. Add approximately 5 g of extracted plant leaves to 0.5 g $MnCO_3$ and about 1 g $Fe(OH)_3$ (wet weight) in graduated or long glass cylinder. Fill cylinder almost to top with water from natural source (iron pipe, lake, pond, stream, etc.). Cover with plastic to prevent evaporation. Incubate for 1–2 weeks in dark at 25°C.

Observation

Look for reddish-brown flakes on sides of cylinder near water surface. Examine flakes in a wet mount under 400–1000 x for filaments of *Sphaerotilus* encrusted with golden-brown layer of iron or manganese oxides. (See Fig. VII-4.) [Reference: M. A. Rouf, and J. L. Stokes. *Arch. Mikrobiol.* **49**, 132–49 (1964).]

48. ENRICHMENT FOR IRON BACTERIA NO. 2

Materials

250-ml Erlenmeyer flask
Mineral medium

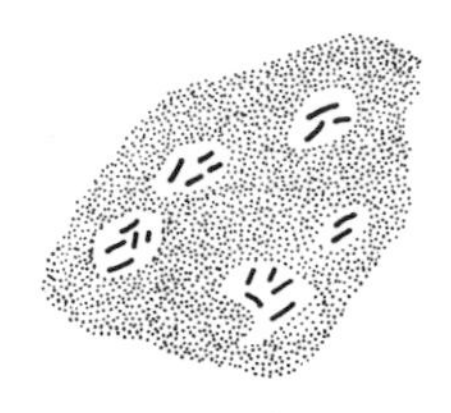

Siderocapsa sp. in ferric hydrate. Short rods or coccid cells.

Sphaerotilus sp. in sheath. Filaments show false branching.

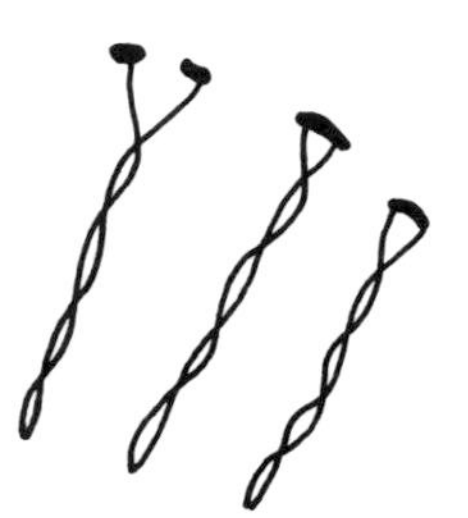

Gallionella sp. Cells at the ends of excretin bands undergoing division.

Leptothrix sp. Cells coming out of ferric hydrate sheath. No branching.

FIG. VII-4. Appearance of some iron bacteria.

$(NH_4)_2SO_4$	0.005 *M*
NaCl	0.005 *M*
KH_2PO_4	0.001 *M*
$MgSO_4 \cdot 7\ H_2O$	0.001 *M*
$CaCO_3$	1.0 g
$Fe(NH_4)_2(SO_4)_2 \cdot 6\ H_2O$	0.6 mg
Water to	100 ml

Procedure

1. Add 100 ml of mineral medium to 250-ml flask.
2. Inoculate with 1.0 g of soil or 5.0 ml of water sample.
3. Cover with plastic to prevent evaporation. Incubate at 25°C in dark for 1-3 weeks.

Observation

Same as for Method No. 47, p. 103. [Reference: Meiklejohn, J. *J. Gen. Microbiol.* 8, 58-65 (1953).]

49. Enrichment for Iron Bacteria No. 3

Materials

250-ml Erlenmeyer flask
Mineral medium

$MgSO_4 \cdot 7\ H_2O$	0.02 g
KH_2PO_4	0.05 g
$FeSO_4 \cdot 7\ H_2O$	2.0 g
Water to	100 ml
pH (adjust with H_2SO_4)	2.5–3.0

Procedure:

See Method No. 48, p. 104.

Observation

See Method No. 47, p. 103.

[Reference: J. W. Beck, and S. R. Elsden. *J. Gen. Microbiol.* **19**, I (1958).]

50. Isolation of Manganese-Oxidizing Microorganisms

Materials

Isolation medium

Yeast extract	0.005 g
$MnSO_4 \cdot 4\ H_2O$	0.002 g
Agar	2.0 g
Water to	100 ml

Potassium metaperiodate

Procedure

1. Grind deposits from pipes.
2. Dilute this deposit, soil, or water sample serially in sterile water and streak on above isolation media agar.

Observation

Manganese-oxidizing microorganisms are recognized by the brown mass of oxidized manganese they produce in agar media. The presence of manganese may be confirmed by oxidizing the brown material to purple permanganate with potassium metaperiodate. They usually belong to the following genera of stalked, budding bacteria: *Hyphomicrobium, Pedomicrobium*; of fungi *Cladosporium, Coniothyrium.*

Enrichment procedure is given in the reference. [Reference: P. A. Tyler, and N. C. Marshall. *Antonie van Leeuwenhoek J. Microbiol. Serol.* **33**, 171–183 (1967).]

51. Enrichment and Isolation of Green Sulfur Bacteria, *Chlorobium* - from H. Larsen J. Bact 64:187-196 (1952)

Materials

Marine or freshwater mud
50-ml glass-stoppered bottles
Nutrient medium*

NH_4Cl	0.1 g
KH_2PO_4	0.1 g
$Na_2S \cdot 9\ H_2O$	0.1 g
$MgCl_2$	0.05 g
$NaHCO_3$	0.2 g
NaCl (for marine organisms)	1.0 g
Water to	100 ml
pH	7.3

Procedure

1. Inoculate 0.1 g mud in sterile medium in 50-ml bottle.
2. Fill bottle to top with medium and close with glass stopper. Avoid trapping air bubble.
3. Incubate in incandescent light at 28°–30°C for 1–3 weeks.
4. Winogradsky column enriched with water containing NH_4Cl (0.1 g/100 ml) and phosphate buffer (0.1 g/100 ml) at pH 7.3 may also be used.
5. Pure cultures may be isolated from shake cultures in nutrient medium solidified with agar (2 g/100 ml). At least 3 successive transfers through shake cultures isolating single microscopically homogeneous colonies are required before the culture is considered pure.

Observation

Look for dark green patches of growth in enrichment and shake cultures. Microscopically the green bacterial cells are gram-negative, nonmotile, ovoid rods, 0.9–1.5 μ long by 0.7 μ wide; no endospore. *Chlorobium limicola* grows rapidly (5–6 days) on sulfide agar; *C.*

*At higher pH or lower sulfide concentration, purple sulfur bacteria are likely to predominate.

thiosulfatophilum grows slowly (several weeks). Cells secrete elementary sulfur in sulfide media. [Reference: H. Larsen. *J. Bacteriol.* **64**, 187-96 (1952).]

52. Enrichment and Isolation of Photosynthetic Bacteria

Materials

Enrichment medium	
$(NH_4)_2SO_4$	0.1 g
K_2HPO_4	0.05 g
$MgSO_4 \cdot 7\ H_2O$	0.02 g
NaCl	0.2 g
$NaHCO_3$	0.5 g
Organic*	0.15-0.2 g
Water to	100 ml
pH (adjust with H_3PO_4)	7.0
Isolation medium	
Yeast autolyzate	0.1 g
$Na_2S \cdot 9\ H_2O$	0.01 g
Agar	1.5 g
Water to	100 ml
pH (adjust with H_3PO_4)	7.0
50-ml glass-stoppered bottles	

Procedure and Observation

Inoculate glass-stoppered bottle with 1-2 ml of surface water or mud-suspension from marine or freshwater swamp and fill remainder of the bottle with the enrichment medium. Avoid trapping air bubble when inserting glass stopper. Incubate 3-15 days in incandescent light at 25°-30°C. After red-, green-, purple-, or brown-colored bacteria appear, bottle may be opened and portion of colored bacterial growth is inoculated into shake culture tubes containing isolation agar medium using successive dilutions. These tubes can be kept anaerobic for long periods by covering the surface with a sterile mixture of equal parts paraffin and paraffin oil. Successive isolations by aseptically picking photosynthetic bacteria out of shake cultures and repeating the shake dilution series several times are necessary before the cultures are likely to be pure. If soft glass or plastic tubes are used,

*Organic includes any one of the following: methanol, ethanol, propanol, butanol, glycerol, polyalcohols, fatty acids (use at 0.05 g/100 ml), hydroxy acids (use at 0.05 g/100 ml), dibasic acids, amines, amino acids, peptone, or yeast extract.

the bottoms may be cut and the agar blown or forced out into a sterile Petri dish for picking colonies of photosynthetic bacteria. [Reference: C. B. van Niel. *Bacteriol. Rev.* **8**, 1–118 (1944).]

53. Enrichment and Isolation of the Stalked Photosynthetic Bacterium, *Rhodomicrobium*

Materials

Enrichment medium

KH_2PO_4	0.136 g
$(NH_4)_2SO_4$	0.05 g
$CaCl \cdot 2\,H_2O$	1.0 mg
$MnSO_4 \cdot 4\,H_2O$	0.25 mg
Na_2HPO_4	0.213 g
$MgSO_4 \cdot 7\,H_2O$	0.02 g
$FeSO_4 \cdot 7\,H_2O$	0.5 mg
$Na_2MoO_4 \cdot 2\,H_2O$	0.25 mg
Agar	1.9 g
Water to	100 ml
pH (adjust with NaOH)	7.2

Isolation medium

KH_2PO_4	0.15 g
$(NH_4)_2SO_4$	0.1 g
NaCl	0.2 g
$MgSO_4 \cdot 7\,H_2O$	0.01 g
Yeast extract	0.1 g
Water to	79.0 ml

Sterilize the preceding solution by filtration and add aseptically:

$NaHCO_3$(5%, w/v)	10.0 ml
$Na_2S \cdot 9\,H_2O$(3%, w/v)	1.0 ml
Sodium acetate (2%, w/v)	10.0 ml
pH (adjust with NaOH)	7.4

1. Fill tall sterile glass beaker (400 ml) with sterile enrichment medium almost to top.
2. Add 1–5 g of soil sample (or 5–10 ml of water sample).
3. Cover top with aluminum foil and incubate on window sill at room temperature.

Observation

After 4–7 weeks, orange-brown layer appears on soil or sides of beaker. This is probably *Rhodomicrobium.* To isolate place small sample of orange-brown material into 500-ml glass-stoppered or screw-cap bottle completely filled with sterile isolation medium. Bottles must be completely filled. Incubate for 3–5 weeks at 25°–30°C in tungsten light. A brownish red turbidity, pellicle, and layer on glass will develop. This culture may be purified by isolating colonies on a shake culture of the isolation medium solidified with 1.8% agar. (See Fig. VII-5.) See anaerobic techniques. [Reference: P. Hirsch, and S. F. Conti. *In* "Anreicherungskultur und Mutantenauslese" (H. G. Schlegel, ed.), pp. 100–110. Fischer Verlag, Stuttgart, 1965.]

54. ENRICHMENT FOR *Rhodomicrobium vannielii*

Materials

Enrichment medium	
$NaHCO_3$	0.5 g
NaCl	0.2 g
$(NH_4)_2SO_4$	0.1 g
K_2HPO_4	0.05 g
$MgSO_4 \cdot 7\ H_2O$	0.01 g
$Na_2S \cdot 9\ H_2O$	0.01 g
Ethanol [Add ethanol aseptically (or propanol, butanol, acetate, propionate, butyrate, valerate, lactate).]	0.2 ml
Water to	100 ml
pH	7.0

Procedure

1. Inoculate mud into glass-stoppered bottle completely filled with above enrichment medium and incubate in the light at 25°–30°C. After 7 days colonies of *R. vannielii* may be the most prominent, although it may be outnumbered by other nonsulfur purple bacteria.
2. Colonies may be isolated and purified from shake cultures containing above medium supplemented with agar (1.5%) and yeast autolyzate (0.2%).

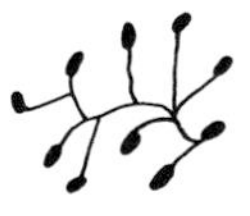

FIG. VII-5. *Rhodomicrobium* sp. Cells connected at ends to stalk.

Observation

Look for salmon-pink to deep orange-red growth. Cells are oval to round and attached by means of a slender branched filament. New cells are initiated by outgrowth of a new filament and swelling of the tip of the filament. Bacteria are nonmotile, nonspore-forming, gram-negative, strictly anaerobic, and nonsulfur-requiring. [Reference: E. Duchow, and H. C. Douglas. *J. Bacteriol.* **58**, 409-16 (1949).]

55. Enrichment and Isolation of Alcohol-Oxidizing Photosynthetic Bacteria

Materials

50-ml glass-stoppered bottles
Nutrient medium*

$(NH_4)_2SO_4$†	0.1 g
KH_2PO_4	0.2 g
$MgCl_2$	0.05 g
$NaHCO_3$	0.4 g
Ethanol (or other alcohol)	2.0 ml
Water to	100 ml
pH	7-8

Procedure

1. Prepare above medium without autoclaving and add it to 60-ml bottles.
2. Add 1.0 g of soil or mud to each bottle.
3. Fill bottle to exclude air bubbles, close with stopper, and incubate at 25°-28°C in incandescent light.
4. Pure cultures may be isolated by using the agar-shake method.

Observation

Photosynthetic sulfur bacterial cultures are red or green while the nonsulfur photosynthetic bacteria will appear as shades of red or brown. Microscopically they may be short or long rods, spirals, or cocci. [Reference: J. W. Foster. *J. Bacteriol.* **47**, 355-72 (1944).]

*Above medium may be supplemented with yeast autolyzate (0.1 g/100 ml) for required growth factors.

†To avoid purple sulfur bacteria which are likely to appear substitute NH_4Cl for $(NH_4)_2SO_4$.

56. Enrichment for Sulfur Bacteria

Materials

Tall glass vessel (glass jar or cylinder)
NaCl
Phosphate buffer at pH 7.0
$CaSO_4$

Procedure

1. Fill tall glass vessel almost half full with a wet slurry of powdered filter paper, calcium sulfate, and the sample (usually mud or polluted water).
2. Add an equal volume of water containing a trace of phosphate at a pH of *ca.* 7.0. This is a Winogradsky column.
3. Marine samples should be diluted with water having a salt content of 2.5%. Brackish water samples should be treated the same way.
4. Incubate column in light (incandescent or sun).

Observation

At the air-water surface sulfur-oxidizing bacteria (*Thiobacillus*) or sulfide-oxidizing (*Beggiatoa* or *Thiothrix*) bacteria will appear. Marine columns yield *Thiovulum* and *Thioplaca.* In the anaerobic zone on the illuminated side from the top down appear rust-colored Athiorhodaceae, e.g., *Rhodopseudomonas* and *Rhodospirillum*; red Thiorhodaceae such as *Chromatium* and *Thiopedia.* Below these at higher H_2S levels, green *Chlorobium* appear. [Reference: J. R. Postgate. *Lab. Pract.* **15**, 1239–44 (1966).

57. Selective Procedures for the Isolation and Growth of Anaerobic Sulfide-Oxidizing Bacteria

Materials

Medium T*

KH_2PO_4	0.1 g
NH_4Cl	0.1 g

*Just before use of above medium sterilize the following solutions and add aseptically: (a) $NaHCO_3$, 0.2%; (b) $Na_2S \cdot 9\ H_2O$; 0.1% for *Chlorobium limicola*; 0.02% for *C. thiosulfatophilum, Chromatium, Thiopedia*; (c) $Na_2S_2O_3 \cdot 5\ H_2O$, 0.1% for *C. thiosulfatophilum, Chromatium, Thiopedia*; (d) Na malate, 0.1% for *Chromatium, Thiopedia*; (e) H_3PO_4 to give: pH of 7–7.2 for *Chlorobium* species; pH 8–8.4 for *Chromatium, Thiopedia.*

$MgCl_2 \cdot 6\ H_2O$	0.05 g
NaCl	1.0 g
Trace elements	0.1 ml
Water to	100 ml

Medium P

Solution 1

$CaCl_2$	0.4 g
Water to	100 ml

Sterilize by autoclaving.

Solution 2a

EDTA	0.5 g
$FeSO_4 \cdot 7\ H_2O$	0.2 g
$ZnSO_4 \cdot 7\ H_2O$	10.0 mg
$MnCl_2 \cdot 4\ H_2O$	3.0 mg
H_3BO_3	30.0 mg
$CoCl_2 \cdot 6\ H_2O$	20.0 mg
$CuCl_2 \cdot 2\ H_2O$	1.0 mg
$NiCl_2 \cdot 6\ H_2O$	2.0 mg
$Na_2MoO_4 \cdot 2\ H_2O$	3.0 mg
Water to	1000 ml

Solution 2b

Vitamin B_{12}	2.0 mg
Water to	100 ml

Solution 2c

KH_2PO_4	1.0 g
KCl	1.0 g
NH_4Cl	1.0 g
$MgCl_2 \cdot 6\ H_2O$	1.0 g

Dissolve solution 2c in 70 ml distilled water plus 30 ml solution 2a plus 3 ml solution 2b.

Solution 3

$NaHCO_3$	4.5 g
Water	900 ml

Saturate solution 3 with CO_2 for 30 minutes by bubbling. Mix with entire solution 2 and filter-sterilize under pressure of CO_2.

Solution 4

$Na_2S \cdot 9\ H_2O$	3.0 g
Water to	200 ml

Sterilize by autoclaving. Neutralize with sterile H_2SO_4 before use.

Mix entire solutions 2 + 3 or appropriate portions with solution 1 in 1:2 ratio, respectively. Add solution 4 to give sulfide concentration for desired bacterium as described for (b) and (c) in footnote, p. 111. Green bacteria do best at pH 6.7–6.9; purple Athiorhodaceae do best at pH 6.9–7.2. Tightly closed filled bottles of medium P may be stored in the dark for several months.

Procedure

1. Incubate bottles with soil or water sample and fill to top with medium T or P.
2. Stopper tightly to exclude air.
3. Incubate in light (incandescent or sun) for 7–14 days at 25°–30°C.

Observation

Green sulfur bacteria *(Chlorobium)* are favored by a pH of 7 and an H_2S concentration of 3 m*M*. Red bacteria (*Chromatium, Thiopedia*) are favored by a pH of 8 and an H_2S concentration of 0.6 m*M*. Medium P is useful for growing Vitamin B_{12}-requiring Thiorhodaceae, eg., *Chromatium okenii, Rhodothiospirillum* spp. as well as *Chlorobium* and *Thiopedia*. [Reference: J. R. Postgate, *Lab. Prac.* **15**, 1239–44 (1966).]

58. Enrichment and Isolation of Sulfur-Oxidizing Bacteria

Material

250-ml flasks (cotton or screw-top)

Enrichment medium

K_2HPO_4	0.1 g
$MgSO_4 \cdot 7\,H_2O$	0.05 g
NH_4NO_3	0.1 g
Sulfur flowers	1.0 g
$CaCO_3$	1.0 g
Water to	100 ml

Isolation medium

(a) Basal medium

K_2HPO_4	0.05 g
$(NH_4)_2SO_4$	0.05 g
$MgSO_4 \cdot 7\,H_2O$	0.025 g
$CaCl_2$	0.01 g
Na_2CO_3	0.01 g
Sodium silicate	0.1 ml

Trace elements No. 1*	1.0 ml
Agar	1.5 g
Water to	100 ml

(b) Polysulfide solution

Saturate a saturated solution of Na_2S in water with elemental sulfur.

Procedure

1. Soil (1.0 g) or water sample (5 ml) is inoculated into 50-100 ml of the enrichment medium in a flask.
2. The flask is then covered with plastic to prevent evaporation.
3. The enrichment flask is incubated for 1-3 weeks at 25°C in the dark.
4. The enrichment may be repeated by transferring fluid from the last flask to a new enrichment flask until bacteria are found in the fluid from the enrichment flask when examined microscopically.
5. Isolation plates are prepared as follows: 25-30 ml of melted, sterile basal medium is poured into a Petri dish containing 1-1.5 ml of a 0.1 *N* HCl solution.
6. Mix well and allow agar to harden.
7. Two milliliters of the polysulfide solution (above) are added to 1000 ml of the basal medium.
8. Ten milliliters of sterile basal medium containing polysulfide is now poured over the previously poured plates and allowed to harden.
9. The HCl diffuses into the top layer and precipitates the sulfur as a fine suspension.
10. Any H_2S that may be formed is removed by drying the plates in a 37°C incubator.
11. These plates are now streaked with an inoculum from the enrichment flask containing bacteria and incubated for 5-7 days at 25°C in dark.

Observation

Clear zones appear around the colonies of bacteria capable of oxidizing sulfur. [Reference: K. T. Wieringa, *Antonie van Leeuwenhoek J. Microbiol. Serol.* **32**, 183-6 (1966).]

*See Tables VI-5,6,7.

59. Selective Procedures for the Isolation and Growth of *Thiobacillus* Species

Materials

Medium S	
Sulfur* (or $Na_2S_2O_3 \cdot 5\ H_2O$)	1.0 g (or 0.5 g)
$(NH_4)_2SO_4$	0.2–0.4 g
KH_2PO_4	0.2–0.4 g
$CaCl_2$	0.025 g
$MgSO_4 \cdot 7\ H_2O$	0.05 g
Trace metals†	1.0 ml
Water to	100 ml
Medium R	
NH_4Cl	0.05 g
$MgCl_2 \cdot 6\ H_2O$	0.05 g
KII_2PO_4‡	0.2 g
$Na_2S_2O_3 \cdot 5\ H_2O$	0.5 g
KNO_3	0.2 g
$NaHCO_3$‡	0.1 g
Trace metals†	1.0 ml
Water to	100 ml
Medium F	
$(NH_4)_2SO_4$	0.015 g
KCl	0.005 g
$MgSO_4 \cdot 7\ H_2O$	0.05 g
KH_2PO_4	0.005 g
$Ca(NO_3)_2 \cdot 4\ H_2O$	0.001 g
$FeSO_4$ (10%)‡	1.0 ml
Trace metals†	1.0 ml
Water to	100 ml

Procedure

1. Use shallow aerobic cultures in conical flasks.
2. Inoculate with soil or water sample.
3. Incubate for 3–7 days at 25°–30°C. Growth is indicated by turbidity and a drop in pH.

*Media with sulfur sterilized by steaming for 30 minutes for 3 successive days.

†Trace metals: Ethylenediaminetetraacetic acid, 5.0 g; $ZnSO_4 \cdot 7\ H_2O$, 2.2 g; CaCl, 0.544 g; $MnCl_2 \cdot 4\ H_2O$, 0.506 g; $FeSO_4 \cdot 7\ H_2O$, 0.499 g; $(NH_4)_6Mo_7O_{24} \cdot 4\ H_2O$, 0.11 g; $CuSO_4 \cdot 5\ H_2O$, 0.157 g; $CoCl_2 \cdot 6\ H_2O$, 0.161 g; Water to 100 ml; pH (adjust with KOH), 6.0.

‡Sterilize these solutions separately and add them aseptically to medium.

Observation

Medium S at initial pH 5.0 selects for the *T. thiooxidans* group while at pH 7.5 the *T. thioparus* group is selected. Medium R selects for the denitrifiers, i.e., *T. ferrooxidans* group at pH 3.5; growth is indicated by the browning of the medium and a lowering of the pH. [Reference: J. R. Postgate, *Lab. Prac.* **15**, 1239–44 (1966).]

60. Enrichment and Isolation of *Thiobacillus*

Materials

250-ml Flask

Thiosulfate-mineral medium*

$Na_2S_2O_3 \cdot 5\,H_2O$	1.0 g
NH_4Cl	0.1 g
KH_2PO_4	0.1 g
$MgCl_2 \cdot 6\,H_2O$	0.05 g
Water to	100 ml
pH	6.8

Procedure

1. Add 1.0 g of mud or soil to 100 ml of above medium in a 250-ml flask.
2. Incubate in dark at 30°C for 1–2 weeks.

Observation

Thiobacillus thioparus multiplies first and produces enough acid for the multiplication of *T. thiooxidans*. Pure cultures may be isolated by streaking fluid from flask onto surface of thiosulfate-mineral medium hardened with agar (1.5 g/100 ml). Facultative autotrophs may be identified by their very slow multiplication (7–10 days) on mineral thiosulphate agar and their rapid growth if medium is supplemented with low concentrations of yeast extract, casein hydrolyzate, etc. [Reference: J. London, *Arch. Mikrobiol.* **46**, 329–37 (1963).]

61. Selective Procedures for the Isolation and Growth of Sulfate-Reducing Bacteria

Materials

Medium B

KH_2PO_4	0.05 g

*May be enriched with yeast extract (0.005 g/100 ml) for isolation of facultative autotrophs.

NH_4Cl	0.1 g
$CaSO_4$	0.1 g
$MgSO_4 \cdot 7\,H_2O$	0.2 g
Sodium lactate	0.35 g
Yeast extract	0.1 g
Ascorbic acid	0.1 g
Thioglycollic acid	0.1 g
$FeSO_4 \cdot 7\,H_2O$	0.5 g
Water* to	100 ml
pH	7.0–7.5
Medium C†	
KH_2PO_4	0.05 g
NH_4Cl	0.1 g
Na_2SO_4	0.45 g
$CaCl_2 \cdot 6\,H_2O$	6.0 mg
$MgSO_4 \cdot 7\,H_2O$	6.0 mg
Sodium lactate	0.6 g
Yeast extract	0.1 g
$FeSO_4 \cdot 7\,H_2O$	0.01 g
Sodium citrate $\cdot$ 2 H_2O	0.03 g
Water* to	100 ml
pH	7.5
Medium D (sterilize by filtration)	
KH_2PO_4	0.05 g
NH_4Cl	0.1 g
$CaCl_2 \cdot 2\,H_2O$	0.01 g
$MgCl_2 \cdot 6\,H_2O$	0.16 g
Yeast extract	0.1 g
$FeSO_4 \cdot 7\,H_2O$	0.01 g
Sodium pyruvate or	0.35 g
choline chloride	0.1 g
Water* to	100 ml
pH	7.5
Medium E (sterilize by autoclave; use before hardening)	
KH_2PO_4	0.05 g
NH_4Cl	0.1 g
Na_2SO_4	0.1 g
$CaCl_2 \cdot 6\,H_2O$	0.1 g

*NaCl (2.5%) must be added or seawater used in place of distilled water for marine and brackish forms.

†Medium may be cloudy after autoclaving but should clear when cool.

$MgCl_2 \cdot 6\ H_2O$	0.2 g
Sodium lactate	0.35 g
Yeast extract	0.1 g
Ascorbic acid	0.1 g
Thioglycollic acid	0.1 g
$FeSO_4 \cdot 7\ H_2O$	0.05 g
Agar	1.5 g
Water* to	100 ml
pH (adjust with NaOH)	7.6

Medium N

Medium C with $(NH_4)_2SO_4$ (0.7 g) in place of Na_2SO_4

Procedure and Observation

Bottle containing sample with medium B is filled to the brim and stoppered to exclude air. Blackening after incubation indicates sulfate-reducing bacteria. *Desulfotamaculum nigrificans* (a spore former) grows only at 55°C. Other species of the *Desulfotamaculatum* are mesophilic (30° or 37°C).

Desulfovibrio species are recognized by collecting 1 ml of culture supernatant (avoid FeS precipitate), which is then centrifuged, and the supernatant discarded. Add 1 drop of 2 *N* NaOH to pellet under long wave UV-light (3750 Å; Woods light). *Desulfovibrio* species give a strong red fluorescence.

Growth in media containing 0.1% NaCl and in the presence of Hibitane (chlorhexidine; 1 mg/ml) but not in freshwater media suggests *Desulfovibrio salexigens*.

Desulfovibrio desulfuricans grows in medium D in the presence of pyruvate or choline and in the absence of sulfate. In choline medium it may be recognized by the smell of trimethylamine resulting from choline catabolism.

Media C, E and N are used primarily for the mass culture of sulfate-reducing bacteria. [Reference: J. R. Postgate, *Lab. Prac.* **15**, 1239–44 (1966).]

62. Isolation of Sulfate-Reducing Bacteria

Materials

Nutrient agar supplemented with Na_2SO_4, 0.5 g/100 ml and $FeSO_4$, 0.005 g/100 ml.

*NaCl (2.5%) must be added or seawater used in place of distilled water for marine and brackish forms.

Procedure

1. Sterilize agar and pour into Petri dishes.
2. Streak surface of plates with soil suspension in water or water sample.
3. Incubate in dark at 25°-30°C *anaerobically* for 1-2 weeks.

Observation

Look for black colonies resulting from precipitation of iron sulfide (Fig. VII-2). Repeat procedure until colonies are homogeneous. [Reference: M. Stephenson, and L. H. Stickland, *Biochem. J.* **25**, 215-20 (1931).]

63. Isolation of Hydrogen Sulfide-Producing Bacteria

Materials

Nutrient medium

Any mineral salts or organic medium may be used if they contain sulfate for mineral media or cystine (0.01 g/100 ml) and Na_2SO_4 (0.05 g/100 ml) for organic media. The medium chosen must contain an iron or lead salt. Ferric ammonium citrate (0.03 g/100 ml) is less toxic than lead acetate (0.03 g/100 ml).

Procedure

1. Sterilize agar medium with iron (or lead salt) as described in materials and prepare plates.
2. Allow to harden and dry.
3. Streak surface of plate with soil, mud, or water sample.
4. Incubate for several days at 20°-37°C (depends on source of sample) in dark.

Observation

Look for colonies containing and surrounded by areas of black precipitate (iron or lead sulfide). (See Fig. VII-2).

64. Isolation of *Desulfovibrio*

Materials

Nutrient medium

Trypticase	1.5 g
Phytone	0.5 g
NaCl	0.5 g

Agar	2.0 g
Water to	100 ml
pH	7.2
Nutrient medium plus salts	
Above medium to	100 ml
$MgSO_4 \cdot 7\,H_2O$	0.2 g
Sodium lactate (60%)	0.4 ml (v/v)
$Fe(NH_4)_2(SO_4)_2 \cdot 6\,H_2O$	0.05 g
pH	7.2-7.4

Hydrogen tank
Hydrogen generator ($H_2SO_4 + Na_2S \cdot 9\,H_2O$)

Procedure

1. Streak soil or water sample on the surface of Petri dishes with either of the above media within 4-5 hours after agar hardens. This minimizes saturation of agar with oxygen.
2. Place Petri dishes in a Brewer jar or dessicator and evacuate air with pump or water aspirator; replace air with hydrogen from tank or generator. Do this three times.
3. Incubate at 20°-25°C for 1-3 weeks.

Observation

Look for pale yellow transparent colonies on nutrient medium. *Desulfovibrio* cells are comma-shaped and nonmotile. In liquid media they are motile. Nutrient medium plus salts appears black at first; bacteria grow well on this medium. Medium turns blue and then white on exposure to air after growth. Well-isolated colonies may be surrounded by dark color while medium is light colored after exposure to air. [Reference: W. P. Iverson, *Appl. Microbiol.* **14**, 529-34 (1966).]

65. Enrichment and Isolation of *Beggiatoa*

Materials

Hay
125-ml Erlenmeyer flasks
Medium

Beef extract	0.2 g
Agar	1.0 g
Water to	100 ml

Procedure

1. Dried hay is extracted 3 times at 100°C in large volumes of water.
2. Extracted hay is drained and dried at 37°C.
3. Enrichment medium is prepared by distributing extracted hay suspension (0.8 g/100 ml) in 70-ml amounts in 125-ml flasks.
4. Sterilize and inoculate with mud from fresh- or saltwater environment.
5. Incubate in the dark at 28°C for 10 days.

Observation

Growth of *Beggiatoa* appears as a white film on the surface of the medium and submerged upper walls of the flask along with a strong H_2S odor. Large numbers of ciliates appear also. *Beggiatoa* may be isolated by streaking the white film onto beef extract agar medium. Pick filaments that move away from the contaminants and reinoculate on the agar medium. Repeat this isolation procedure until cultures appear pure on inoculation into various test media for contaminants (nutrient broth, nutrient agar, sugar broths, etc.). [Reference: H. L. Scotten, and J. L. Stokes, *Arch. Mikrobiol.* **42**, 353-68 (1962).]

66. ENRICHMENT FOR *Thiovulum*

Materials

Algae
Seawater
Jar or tall beaker

Procedure

1. Place decaying algae (*Ulva*) on the bottom of a jar or beaker and slowly trickle seawater in at bottom allowing excess to overflow.
2. *Thiovulum* grow in long strands or veils in areas where proper concentrations of O_2 and H_2S exist.
3. The strands may be transferred to a similar container minus algae if H_2S is added once or twice daily.

Observation

Thiovulum grows in strands. The cells are oval, gram-negative, and contain sulfur granules. [Reference: J. W. M. La Rivière, *In* "Symposium on Marine Microbiology" (C. H. Oppenheimer, ed.), pp. 61-72. Thomas, Springfield, Illinois, 1963.]

67. Enrichment and Isolation of Marine Ammonia-Oxidizing Bacteria

Materials

Water sample
500-ml flask
Nutrient medium ($CaCO_3$ 0.1 g included in solid media)

NaCl	2.5 g
$(NH_4)_2SO_4$	0.3 g
K_2HPO_4	0.1 g
$MgSO_4 \cdot 7\ H_2O$	0.1 g
Chelated iron ("Atlas EDTA")	0.01 mg
Water to	100 ml
pH (adjusted after sterilization with 1 *N* NaOH)	7.5

Procedure

1. Add 100 ml of water sample to a 500-ml flask containing 1 μM K_2HPO_4 and 50 μM $(NH_4)_2SO_4$.
2. Flask is shaken for 60–90 days on a mechanical shaker at 20°C.
3. Samples periodically tested for the presence of nitrite (see Method No. 72). If nitrite is present, a sample from flask is inoculated into nutrient medium. As ammonia is oxidized more is added aseptically.

Observation

Bacteria growing on above medium where the only energy source is the oxidation of NH_4^+ and production of nitrite are ammonia-oxidizing bacteria. [Reference: S. W. Watson, *In* "Symposium on Marine Microbiology" (C. H. Oppenheimer, ed.), pp. 73–84. Thomas, Springfield, Illinois, 1963.]

68. Enrichment and Isolation of Nitrogen-Fixing Bacteria, *Azotobacter*

Materials

250-ml flask (cotton or screw-top)
Nutrient medium

Mannitol*	2.0 g
K_2HPO_4	0.02 g

*May be replaced with calcium malate 0.5 g or sodium propionate 0.5 g.

Water to	100 ml
pH	7.3-7.6

Agar 2.0 g/100 may be added to solidify nutrient medium.

Procedure:

1. Place 50 ml of nutrient medium in flask and inoculate with 0.1-0.2 g of fresh soil.
2. Incubate at 25°-30°C in dark until a film develops on surface.

Observation

Prepare wet mount or smear from surface film and examine under microscope (400-1000 ×) for short thick rods with rounded ends, $4 \times 5 - 7\ \mu$. Pure culture may be isolated by streaking material from the surface film in the flask onto surface of solidified nutrient medium and treating as described above. Only nitrogen-fixing bacteria can grow without a source of nitrogen. [Reference: M. W. Beijerinck, *Zentr. Bakteriol. Parasitenk.* **7**, 561-82 (1901).]

69. Enrichment and Isolation of Nitrogen-Fixing Bacteria

Materials

250-ml flasks (cotton or screw-top)

Double batch

Basal medium

K_2HPO_4	0.1 g
$MgSO_4 \cdot 7\,H_2O$	0.02 g
$CaCO_3$	0.05 g
$FeCl_3 \cdot 6\,H_2O$	0.025 g
$NaMoO_4 \cdot 2\,H_2O$	0.05 g
Water to	100 ml

Procedure

1. Dispense 80 ml in each flask.
2. Sterilize.
3. Prepare a sterile glucose or sucrose solution (10%) to be added aseptically to basal medium for a final concentration of 1%. Rhamnose (1%) may be used to enrich selectively for *Azotobacter vinelandii.* Starch used without autoclaving and added to basal medium will selectively support *Azotobacter chroococcum.* Sugar plus basal medium at a pH of 5.0-6.0 will select for the growth of *Azotobacter macrocytogenes* and *Beijerinckia.*
4. Inoculate flask with 1 g of soil, 20 ml of water, or 1 g of plant

material; incubate at room temperature in dark. Shake cultures are more efficient than stationary cultures.
5. After growth appears transfer an aliquot to a fresh flask; the growth in original flask includes nonnitrogen-fixing bacteria.

Observation

Bacteria growing in the absence of inorganic or organic nitrogen are nitrogen-fixing bacteria. These may be isolated and purified by streaking on the surface of the aforementioned media solidified with agar (1.6%). [Reference: Personal communication from Professor D. B. Johnstone, 1968 Department of Agricultural Biochemistry, University of Vermont, Burlington, Vermont.]

70. Isolation of Marine Denitrifying Bacteria

Materials

Isolation medium

KNO_3	0.5 g
Meat extract	0.3 g
Peptone	0.5 g
Agar	1.5 g
Aged seawater to	100 ml
pH	7.0

Reagents for testing for $No_3^- \rightarrow NO_2^- \rightarrow N_2 \rightarrow NH_3$. (See Method No. 72)

See V. B. D. Skerman, "A Guide to the Identification of the Genera of Bacteria." Williams & Wilkins, Baltimore, Maryland, 1959; or "Manual of Methods for Pure Culture Study of Bacteria." Biotech Publ. Geneva, New York, 1955.

Procedure

1. Streak 1.0-ml sample of seawater on surface of isolation medium on Petri dishes.
2. Invert and incubate at 25°–30°C in dark for 48–72 hours.
3. Isolate individual colonies and inoculate them into isolation medium minus agar.
4. Incubate for 48 hours as above.

Observation

Use 1.0-ml aliquot of 48-hour liquid culture to test for the conversion of nitrate to nitrite by the NO_3 reduction test. Test for the conversion of nitrate to ammonia by the tests in Method 72. Test for presence of nitrate by the reduction of nitrate to nitrite with zinc dust followed by the nitrite test above. [Reference: A. Sreenivasan, and R. Venkataraman, *J. Gen. Microbiol.* **15**, 241–47 (1956).]

71. Enrichment for Nitrifying Bacteria

Materials

250-ml Erlenmeyer flask
Mineral medium

$(NH_4)_2SO_4$	0.005 *M*
NaCl	0.005 *M*
KH_2PO_4	0.001 *M*
$MgSO_4 \cdot 7\,H_2O$	0.001 *M*
$CaCO_3$	1.0 g
Water to	100 ml

Procedure

1. Add 50 ml of mineral medium to 250-ml flask.
2. Inoculate with 1.0 g of soil or 5.0 ml of water sample.
3. Cover with plastic to prevent evaporation.
4. Incubate at 25°C in dark for 1–3 weeks.

Observation

Examine liquid near surface for bacteria. If bacteria are present they may be isolated by streaking bacteria on surface of mineral medium solidified with 1.5% agar. Presence of bacteria capable of converting NH_4^+ to NO_2^- or NO_3^- should be confirmed by testing for the presence of nitrite or nitrate with appropriate colorimetric test. [See Method 72 or V. B. D. Skerman, "A Guide to the Identification of the Genera of Bacteria." Williams & Wilkins, Baltimore, Maryland, 1959; or "Manual of Methods for Pure Culture Study of Bacteria." Biotech Publ., Geneva, New York, 1955. Reference: J. Meiklejohn, *J. Gen. Microbiol.* 8, 58–65 (1953).]

72. Qualitative Nitrate to Nitrite to Ammonia Tests

Materials

Nitrite reagent
- A. Sulfanilic acid, 8 g, dissolved in 1 liter 5 *N* acetic acid
- B. 6 ml of dimethyl-α-naphthylamine in 1 liter of 5 *N* acetic acid

Ammonia reagent (Nessler's reagent)
Zinc dust

Procedure and Observations

Nitrate may be reduced to nitrite by microorganisms; both nitrate and nitrite may be reduced to ammonia and/or molecular nitrogen.

1. Nitrite test. Add 1.0 ml of reagent A to 5.0 ml of culture fluid containing nitrate and to 5.0 ml of uninoculated culture medium containing nitrate. Shake well. Add 1.0 ml of reagent B to each 5.0-ml sample. If nitrate is converted to nitrite, a red color is produced. The uninoculated sample should be lighter in color or free of color.

2. Ammonia test. Add 1.0 ml of Nessler's reagent to 5.0 ml of uninoculated culture medium containing nitrate. If nitrate and/or nitrite has been reduced to ammonia, a yellow-orange color will appear. Uninoculated sample should be lighter or free of red color.

3. Nitrate test. To determine if the nitrate is still present unchanged in the presence of negative nitrite and ammonia tests, add a pinch of zinc dust to a new untreated 5.0-ml sample of culture fluid. Shake well and wait 10 minutes; shake occasionally. Add 1.0 ml of nitrite reagents A and B in order. If fluid turns pink to red it indicates that nitrate was present and the zinc dust reduced the nitrate to nitrite.

If the nitrate, nitrite, and ammonia tests are negative, then it is likely that all the nitrate was converted to nitrogen. See tabulation below:

If all the NO_3^- converted to	Nitrite	Ammonia	Nitrate
NO_2^-	+	−	−
NH_4^+	−	+	−
N_2	−	−	−
Unchanged	−	−	+

Quantitative tests for nitrate reduction may be found in the reference list in Method 71.

I. Luminescent Bacteria

73. Isolation of Luminescent Bacteria

Materials

Nutrient agar medium
Gelysate (BBL) 0.5 g

Yeast extract	0.01 g
$FePO_4$	0.01 g
Agar	1.8 g
Seawater to	100 ml

Quebec bacterial colony counter fitted with 10-watt red light bulb and connected to a rheostat

Marine fish or water

Procedure

1. Streak surface of nutrient agar medium in Petri dishes with seawater or other marine sample (mucous from fish).
2. Incubate for 5-7 days at 20°C.

Observation

Examine surface of agar in a darkened area with red illumination reduced to the point that nonluminescent colonies are barely visible. The luminescing colonies will have a bluish tinge. These may be marked and subsequently transferred to fresh nutrient agar medium for purification and subculture. [Reference: B. J. Cosenza, and J. D. Buck, *Appl. Microbiol.* **14**, 692 (1966).]

74. Isolation of Luminous Bacteria *Photobacterium*

Materials

Nutrient medium A	
Agar	1.0 g
Peptone	1.0 g
NaCl	0.5 g
Water to	100 ml
pH	7.6
Nutrient medium B	
Agar	1.0 g
Peptone	1.0 g
Water	25 ml
Aged seawater	75 ml
pH	7.6
Nutrient medium C	
Agar	1.0 g
Peptone	1.0 g
Water	25 ml
Aged seawater	75 ml
pH	7.6

Procedure

1. Streak surface of sterile Petri dishes containing any one of the above media with marine soil or mud. The best source is the gut content or surface slime of marine fish.
2. Incubate Petri dish in the dark at 20–25°C.

Observation

After 1 week of incubation examine agar plates for luminous colonies in a completely blacked out area. Examine plates for several weeks before discarding them. Pure cultures may be isolated by picking luminous colonies and streaking them onto surface of fresh sterile agar. [Reference: R. Spencer, *J. Gen. Microbiol.* **13**, 111–18 (1955).]

J. Budding and Stalked Bacteria

75. ENRICHMENT FOR BUDDING BACTERIA, *Hyphomicrobium*

Materials

250-ml flasks
Soil
Cotton
Mineral medium

$Na_2HPO_4 \cdot 7\ H_2O$	0.02 g
KNO_3	0.04 g
$MgSO_4 \cdot 7\ H_2O$	0.48 mg
$MnCl_2 \cdot 4\ H_2O$	0.01 mg
$FeCl_3 \cdot 6\ H_2O$	0.02 mg
Water to	100 ml
pH (adjust with HCl or NaOH)	7.2

Procedure

1. Mineral medium (50–100 ml) is sterilized in cotton-plugged (or screw-top) 250-ml flasks.
2. Each flask is inoculated with 0.5 g of soil or 1.0 ml of liquid sample.
3. Incubate 4–8 weeks in the dark at 25°C.

Observation

After 10 days look for a thin, colorless pellicle which thickens with time. Make a wet mount and examine the pellicle for budding bacteria under microscope 400-1000 ×. Cells may be stained by making a smear of the pellicle, heat-fix, and then stain with carbol fuchsin for 3 minutes. Examine smear under oil immersion. (See Fig. VII-6). [Reference: P. Hirsch, and S. F. Conti, *Arch. Mikrobiol.* **48**, 339-57 (1964).]

76. Enrichment and Isolation of the Stalked Bacteria *Caulobacter* and *Hyphomicrobium*

Materials

Enrichment medium

KH_2PO_4	0.136 g
$(NH_4)_2SO_4$	0.05 g
$CaCl_2 \cdot 2\,H_2O$	1.0 mg
$MnSO_4 \cdot 7\,H_2O$	0.25 mg
Na_2HPO_4	0.213 g
$MgSO_4 \cdot 7\,H_2O$	0.02 g
$FeSO_4 \cdot 7\,H_2O$	0.5 mg
$Na_2MoO_4 \cdot 2\,H_2O$	0.25 mg
Water to	100 ml
pH (adjust with NaOH)	7.2
(agar 1.9% may be used to solidify)	

Isolation medium for *Caulobacter*

Peptone	1.0 g
Yeast extract	0.3 g
Agar	1.8 g
Water to	100 ml
pH (adjust with NaOH)	7.2

Isolation medium for *Hyphomicrobium*

Enrichment medium plus methylamine · HCl, 0.675%

Procedure

1. Fill tall sterile glass beaker (400 ml) with sterile enrichment medium almost to top.
2. Add 1–5 g of soil sample (or 5–10 ml of water sample.)
3. Cover with aluminum foil.
4. Incubate in dark at 30°C.

FIG. VII-6. *Hyphomicrobium* sp. Cells connected at ends to stalk.

Observation

After 1-4 weeks a pellicle may form. Streak samples of this pellicle on *Caulobacter* isolation agar medium and incubate in dark at 30°C. Look for raised faintly gray or brown colonies which are smooth and glistening. Streak samples of pellicle on *Hyphomicrobium* isolation agar medium. Incubate in dark at 30°C. After 3-8 days look for cream or brown colored colonies with slightly wrinkled surface. [Reference: P. Hirsch, and S. F. Conti, *in* "Anreicherungskultur und Mutantenauslese" (H. G. Schlegel, ed.), pp. 100-110. Fischer Verlag, Stuttgart, 1965.]

77. Isolation of *Caulobacter* from Water

Materials

Enrichment medium	
Peptone	0.005-0.01 g
Water to	100 ml
Isolation medium (freshwater)	
Peptone	0.1 g
Yeast extract	0.05 g
$MgSO_4 \cdot 7\ H_2O$	0.01 g
Agar	1.0 g
Water to	100 ml
Isolation medium (marine)	
Peptone	0.05 g
Casein hydrolyzate	0.05 g
Agar	1.0 g
Seawater (or NaCl, 3%) to	100 ml

Procedure

1. Allow enrichment medium, made with 100 ml of unsterilized water sample, to incubate at room temperature (13°-19°C for marine forms) in a beaker in the dark or light until a pellicle develops.
2. Cover but do not seal container.

Observation

Examine pellicle for stalked bacteria. These may be purified by streaking material from the pellicle on surface of isolation agar medium in Petri dishes. Look for colonies containing stalked bacteria. Purified cultures may require addition of 0.1 $\mu\mu$g of riboflavine per ml for maintenance in isolation medium. [Reference: J. L. Stove,

In "Anreicherungskulkur und Mutantenauslese" (H. G. Schlegel, ed.), pp. 95-98, Fischer Verlag, Stuttgart, 1965.]

K. Spiral Bacteria

78. Isolation of Freshwater Spiral Bacteria, *Spirillum*

Materials

Cellulose ester filter discs (Millipore) with a diameter of 47 mm and an average pore size of 0.45 μ

Nutrient medium

Peptone	0.5 g
Yeast extract	0.05 g
Tween 80	0.002 g
K_2HPO_4	0.01 g
Agar	1.0 g
Water to	100 ml
pH	7.2

Procedure

1. Sterilize cellulose filter discs in an empty Petri dish.
2. Sterilize nutrient agar medium and prepare agar plates.
3. When agar plates are firm, place one cellulose filter disc on surface of each plate.
4. Place 0.05 ml of pond or stream water in center of filter disc, incubate 2-5 hours at room temperature, and then remove disc.
5. Incubate plates 3 or more days at 25°C in dark.

Observation

Look for spreading, semitransparent areas of growth within agar medium (Fig. VII-7). The subsurface material should show thin spiral bacteria in a wet mount examined under microscope (400-1000 ×). Sometimes other small motile bacteria may make their way through the filter but they usually form small surface colonies. [Reference: E. Canale-Parola, S. L. Rosenthal, and D. G. Kupfer, 1966. *Antonie Van Leeuwenhoek J. Microbiol. Serol.* **32**, 113-24 (1966).]

79. Isolation of Marine Halophilic Spirilla

Materials

Fresh marine shellfish

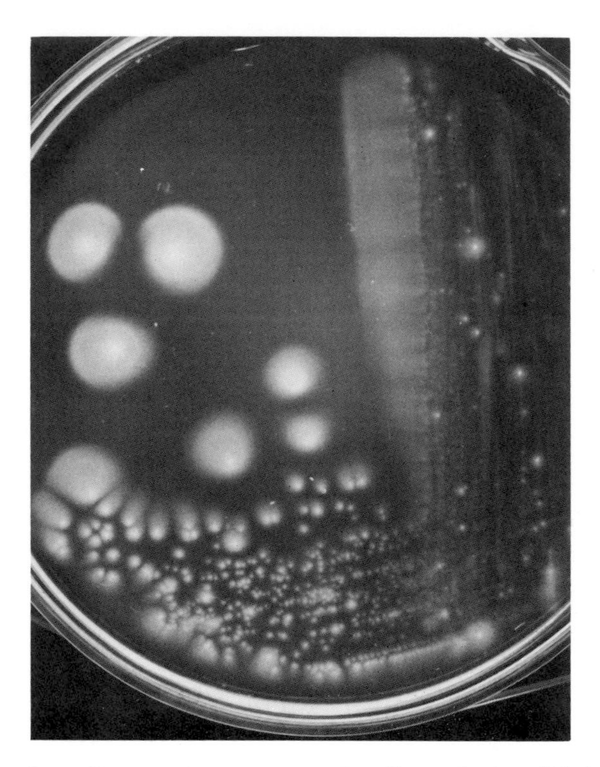

FIG. VII-7a. Spreading, semitransparent, subsurface colonies of *Spirillum*. [From Canale-Parola, E., Rosenthal, S. L., and Kupfer, D. G., *Antonie van Leeuwenhoek J. Microbial. Serol.* **32**, 113-24 (1966).]

Nutrient medium A	
NaCl	2.5 g
$MgCl_2$	0.3 g
$MgSO_4 \cdot 7\ H_2O$	0.1 g
$CaSO_4$	0.1 g
K_2SO_4	0.1 g
$CaCO_3$	0.05 g
Peptone	0.25 g
Water to	100 ml
pH	7.0-7.2
Nutrient medium B	
NaCl	2.5 g
Calcium lactate	1.0 g
NH_4Cl	0.1 g
K_2HPO_4	0.05 g

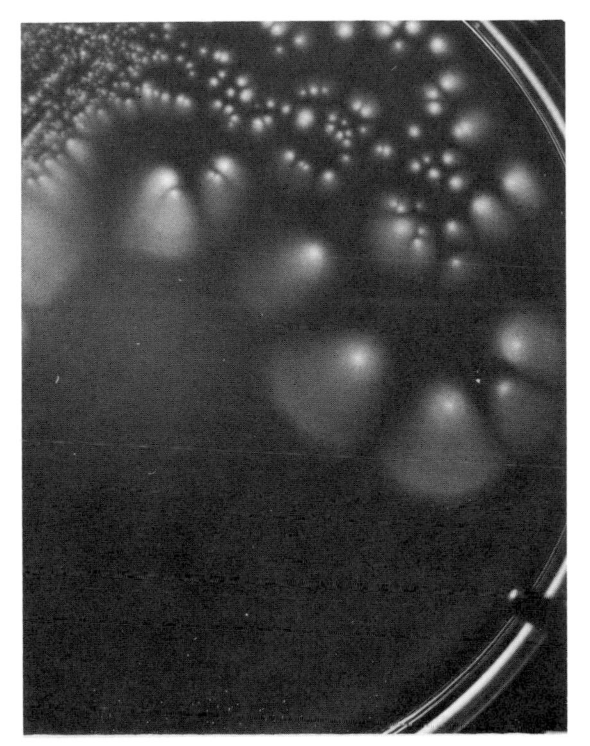

FIG. VII-7b.

$MgSO_4 \cdot 7\ H_2O$	0.05 g
Peptone	0.25 g
Water to	100 ml
pH	7.0–7.2

NaCl solution (2%)

Procedure

1. Place 2–3 well-washed living marine shellfish (snails, clams, etc.) in a Petri dish containing NaCl (2%).
2. Allow shellfish to die and putrify at room temperature.
3. Examine liquid in Petri dish microscopically for spirilla.
4. If spirilla are present, inoculate fluid into a tube containing 5 ml of nutrient medium A (or 1% peptone in seawater) and incubate for 7 days at 20°–23°C.
5. After incubation streak sample of liquid showing spirilla on surface of agar plates containing medium B hardened with agar (1–1.5%).
6. Incubate for 4–5 days at 20°–30°C.

Observation

Look for coarse granular colonies of spirilla with calcium crystals appearing like oil droplets in the colonies under low magnification 100–400 ×). [Reference: N. Watanabe, *Botan. Mag.* (*Tokyo*) **72**, 77–86 (1959).]

80. Isolation of Aquatic Heterotrophic Spirilla

Materials

Enrichment medium*	
Calcium malate (or lactate)	2.0 g
K_2HPO_4	0.05 g
$MgSO_4 \cdot 7\ H_2O$	0.05 g
Peptone	0.01 g
Yeast Extract	0.001 g
Water to	100 ml
pH	7.4
Isolation medium	
Enrichment medium	100 ml
Peptone	0.1 g
Yeast extract	0.01 g
Asparagine	0.05 g
Agar	1.8 g

Procedure

1. Use 125-ml Erlenmeyer flask containing 20 ml enrichment medium.
2. Add 20 ml of water sample or pieces of aquatic plants.
3. Incubate in dark at 30°C for 2–5 days.
4. Look for appearance of a pellicle containing $CaCO_3$ crystals.

Observation

Examine liquid for spirilla. A drop of enrichment fluid containing spirilla may be streaked on isolation medium in Petri dishes. Incubate and examine colonies for presence of spirilla. [Reference: H. W. Jannasch, *in* "Anreicherungskultur and Mutantenauslese" (H. G. Schlegel, ed.), pp. 198–203. Fischer Verlag, Stuttgart, 1965.]

*For marine media use Calcium malate and seawater at pH 7.8.

L. Actinomycetes

81. Media for the Isolation of Actinomycetes

Materials

Czapek's agar	
Sucrose (or glycerol or glucose)	3.0 g
$NaNO_3$	0.2 g
K_2HPO_4	0.1 g
$MgSO_4 \cdot 7\ H_2O$	0.05 g
KCl	0.05 g
$FeSO_4 \cdot 7\ H_2O$	1.0 mg
Agar	1.5 g
Water to	100 ml
pH	7.0-7.3
Glucose–asparagine agar	
Glucose	1.0 g
Asparagine	0.05 g
K_2HPO_4	0.05 g
Agar	1.5 g
Water to	100 ml
pH	6.8
Glycerol–asparagine agar	
Glycerol	3.5 ml
NaCl	0.5 g
$CaCl_2$	0.01 g
$MgSO_4 \cdot 7\ H_2O$	0.03 g
K_2HPO_4	0.25 g
NH_4 lactate	0.65 g
Sodium asparagine	0.35 g
Agar	2.0 g
Water to	100 ml
Starch agar	
Soluble starch	1.0 g
$NaNO_3$	0.1 g
K_2HPO_4	0.03 g
NaCl	0.05 g
$MgCO_3$	0.1 g
Agar	1.5 g
Water to	100 ml
Potato–glucose agar	
Peeled potatoes	20–30 g
Glucose	0.5 g

Agar	2.0 g
Water to	100 ml
pH	6.8–7.2
Oatmeal agar	
Rolled oats	6.5 g
Water to	100 ml

Cook oats to thin gruel in double boiler. Filter through several layers of cheesecloth and bring volume to 100 ml while hot. Add agar (2.0 g). Sterilize.

Soil extract agar	
Beef extract	0.3 g
Peptone	0.5 g
Agar	1.5 g
Soil extract* to	100 ml
pH	7.0

[Reference: S. A. Waksman, *Bacteriol. Rev.* **21**, 1–29 (1957).]

82. Isolation of Soil Actinomycetes

Materials

Nutrient medium	
Glycerol	2.0 g
L-Arginine	0.25 g
NaCl	0.1 g
$CaCO_3$	0.01 g
$FeSO_4 \cdot 7\,H_2O$	0.01 g
$MgSO_4 \cdot 7\,H_2O$	0.01 g
Agar	2.0 g
Water to	100 ml
Antifungal agents	
Pimaricin	50 μg/ml
Cycloheximide	50 μg/ml
Nystatin	50 μg/ml

Procedure

1. Pour 15 ml of sterile nutrient agar medium in Petri dish and allow to harden.
2. Suspend soil sample in sterile water.

*Heat 1 kilogram of garden soil in 2.5 liters of water for 1 hour in an autoclave at 15 psi. Filter and use. Diatomaceous earth may be used to clarify extract if necessary.

3. Dilute and add 1 ml of diluted sample to 14 ml of melted sterile nutrient agar and keep at 48°C in a water bath.
4. Pour 5-ml portions on surface of a Petri dish containing hardened nutrient agar medium.

Observation

Examine agar surface for characteristic mycelial growth. Mycelium branches in young cultures and breaks up into rods and cocci in older cultures. [Reference: J. N. Porter, J. J. Wilhelm, and H. D. Tresner, *Appl. Microbiol.* **8**, 174–178 (1960).]

83. Isolation of Actinomycetes

Materials

Glucose–yeast extract agar (GYE)

Glucose	1.0 g
Yeast extract	1.0 g
Agar	2.5 g
Water to	100 ml
pH	6.8

Procedure

1. Dry GYE plates at 37°C overnite.
2. Inoculate a soil suspension in sterile water over agar surface.
3. Heat in a hot-air oven for 10 minutes at 110°C. (Agar begins to melt if kept longer than 10 minutes.)
4. Incubate plates at 30°C for 5–7 days.

Observation

This procedure inhibits the growth and spread of bacterial and fungal colonies and allows actinomycete colonies to appear. Actinomycetes form mycelia in young cultures; cells may appear filamentous and branching. Filaments break up into rods in older cultures. (See Fig. VII-8). [Reference: A. D. Agate, and J. V. Bhat, *Antonie Van Leeuwenhoek J. Microbiol. Ser.* **29**, 297–304 (1963).]

84. Isolation of Aerobic Actinomycetes

Materials

AGS medium

Arginine · HCl	0.1 g
Glycerol	1.25 g
K_2HPO_4	0.1 g

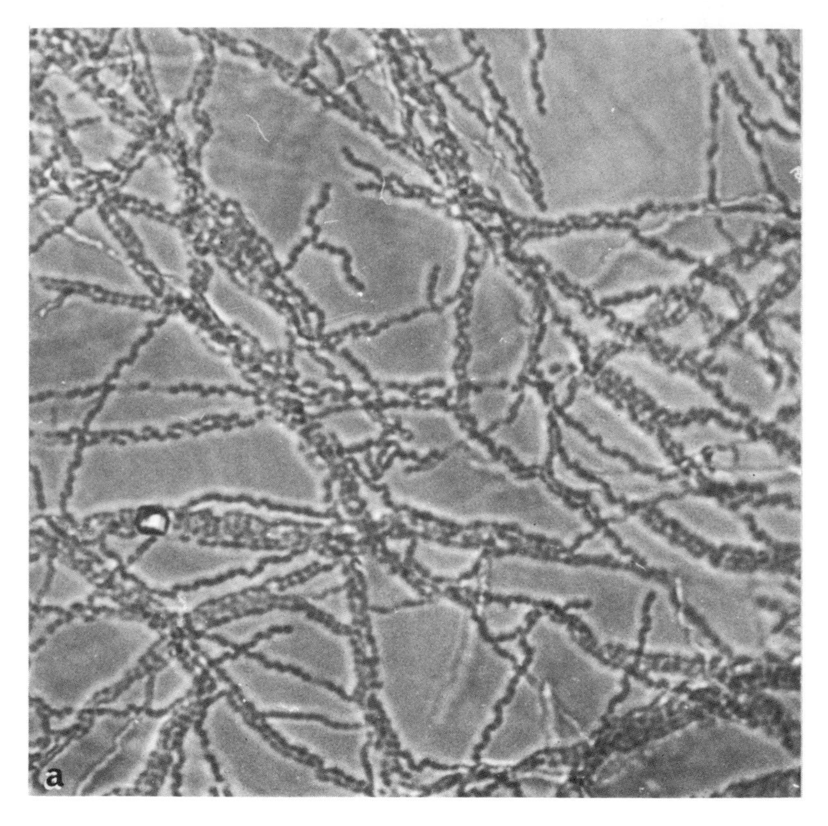

FIG. VII-8a. Illustrations of some actinomycete genera. [From Williams, S. T., Davies, F. L., and Cross, T., *In* "Identification Methods for Microbiologists." (B. M. Gibbs and F. A. Skinner, Eds.), pp. 111-124, Academic Press, New York, 1968.] (a). *Nocardia* type *madurae*, substrate mycelium (x500).

NaCl	0.1 g
$MgSO_4 \cdot 7\ H_2O$	0.05 g
$Fe_2(SO_4)_3 \cdot 6\ H_2O$	1.0 mg
$CuSO_4 \cdot 5\ H_2O$	0.1 mg
$ZnSO_4 \cdot 7\ H_2O$	0.1 mg
$MnSO_4 \cdot H_2O$	0.1 mg
Agar	1.5 g
Water to	100 ml
pH	6.9-7.1

Procedure

1. Air-dried soil (1.0 g) is mixed in a mortar with calcium carbonate (1.0 g).

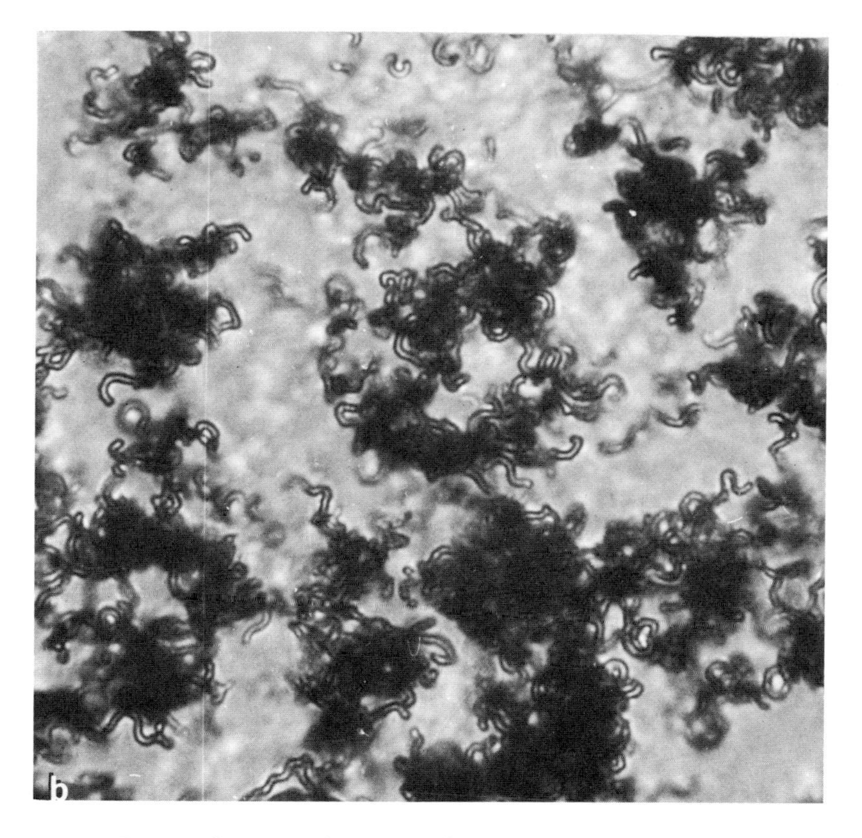

FIG. VII-8b. *Nocardia* type *madurae*, aerial mycelium (x800).

2. Incubate mixture for 10 days in a closed inverted sterile Petri dish in which a high humidity is maintained by water-saturated filter paper discs.
3. Suspend mixture in 100 ml of sterile water in a sterile blender. Mix.
4. Allow sediment to settle for 30 minutes.
5. Dilute supernatant and streak aliquots of the dilutions on AGS medium.
6. Incubate plates for 10 days at 28°C.

Observation

Examine agar surface for characteristic mycelial growth. Mycelium branches in young cultures and breaks up into rods and cocci in older

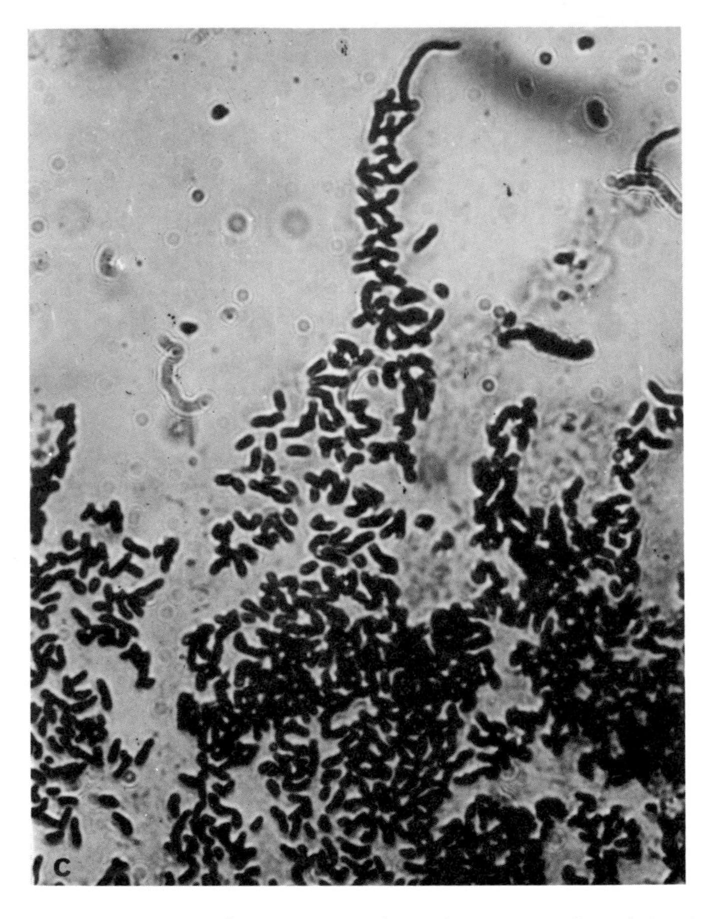

FIG. VII-8c. *Nocardia* type *asteroides*, substrate mycelium (x1000).

cultures. [Reference: M. A. El-Nakeeb, and H. A. Lechevalier, *Appl. Microbiol.* **11**, 75–77 (1963).]

85. Isolation of Marine or Terrestrial Actinomycetes

Materials

Nutrient medium

Soluble starch	1.0 g

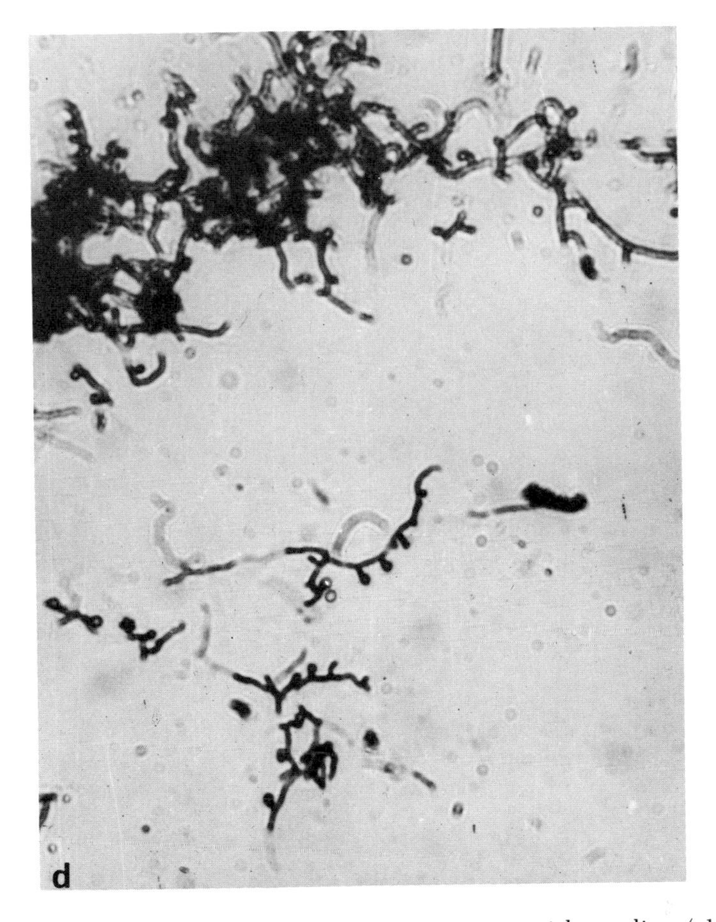

FIG. VII-8d. *Thermoactinomyces vulgaris*, spores on aerial mycelium (x1000).

Casein (dissolved in NaOH)	0.1 g
Agar	1.5 g
Seawater (or distilled water) to	100 ml
pH	7.0-7.5

Procedure

1. Streak 1.0 ml of soil or mud suspended in water (or seawater) on surface of nutrient agar Petri dish.
2. Invert and incubate at 25°-30°C for 1 week.

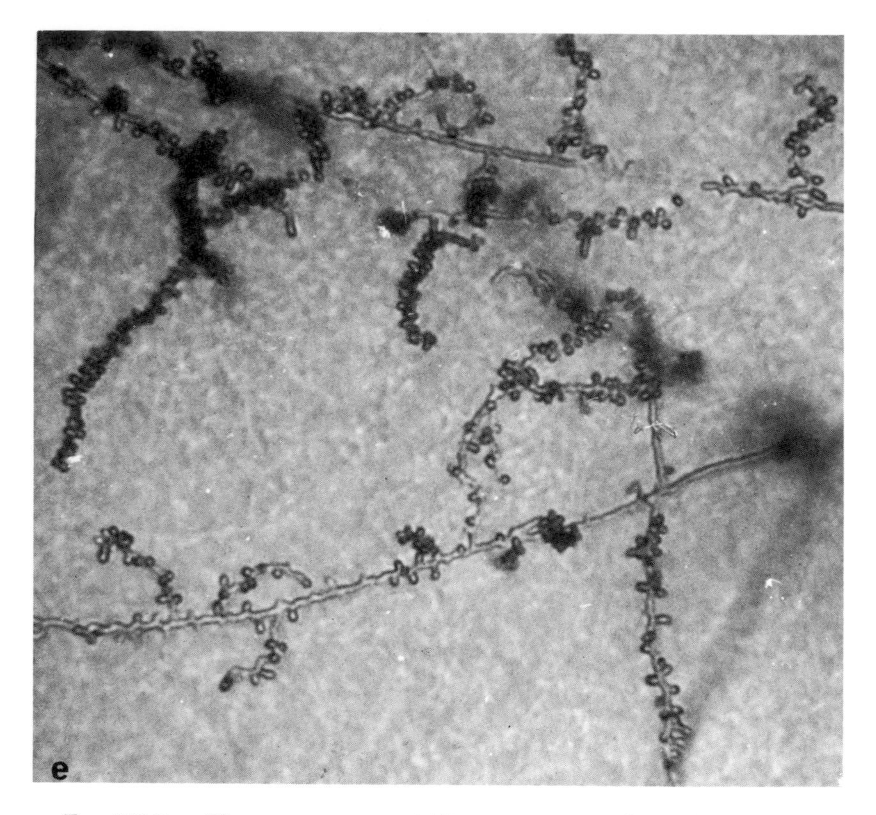

FIG. VII-8e. *Thermomonospora viridis*, spores on aerial mycelium (x800).

Observation

Examine surface for characteristic mycelial growth. Under the microscope mycelial growth is branching in young cultures; mycelium breaks up into rods and cocci as it ages. Mycelium may or may not produce conidia (spore-producing structures at end of filament). (See Fig. VII-7.) [Reference: A. Grein, and S. P. Meyers, *J. Bacteriol.* **76**, 457-63 (1958).]

86. ISOLATION OF STREPTOMYCETES

Materials

Medium

Glycerol (or starch) 1.0 g

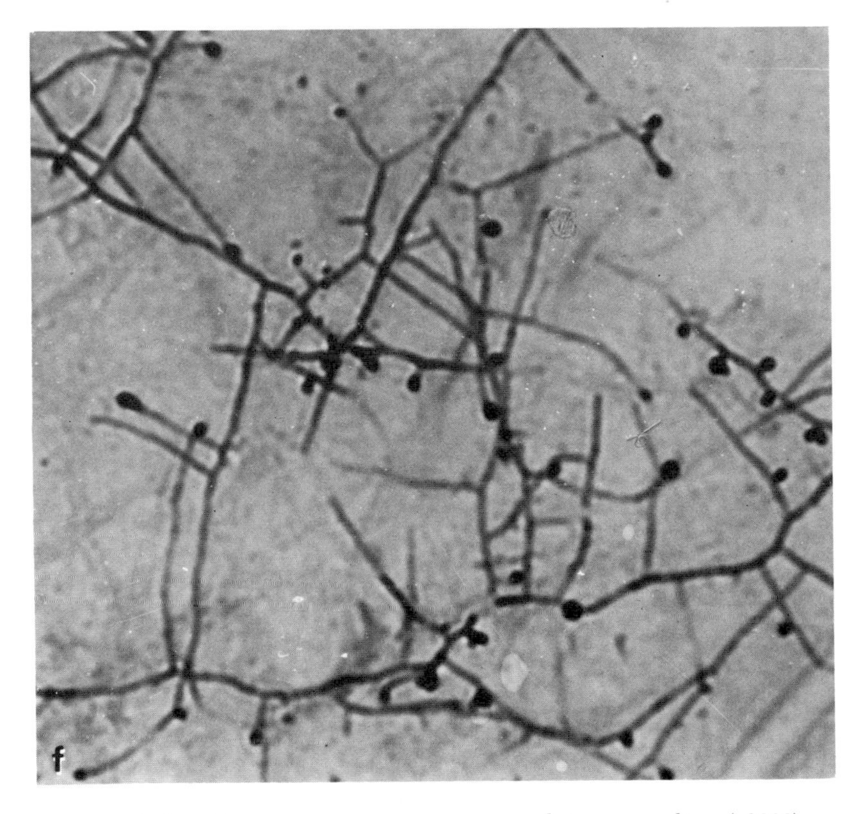

FIG. VII-8f. *Micromonospora* sp., spores on substrate mycelium (x2000).

Casein (vitamin-free)	0.03 g
KNO_3	0.2 g
NaCl	0.2 g
K_2HPO_4	0.2 g
$MgSO_4 \cdot 7\ H_2O$	5.0 mg
$CaCO_3$	2.0 mg
$FeSO_4 \cdot 7\ H_2O$	1.0 mg
Agar	1.8 g
Water to	100 ml
pH	7.0-7.2

Procedure

1. Prepare dilution plates with above medium and soil as the inoculum.

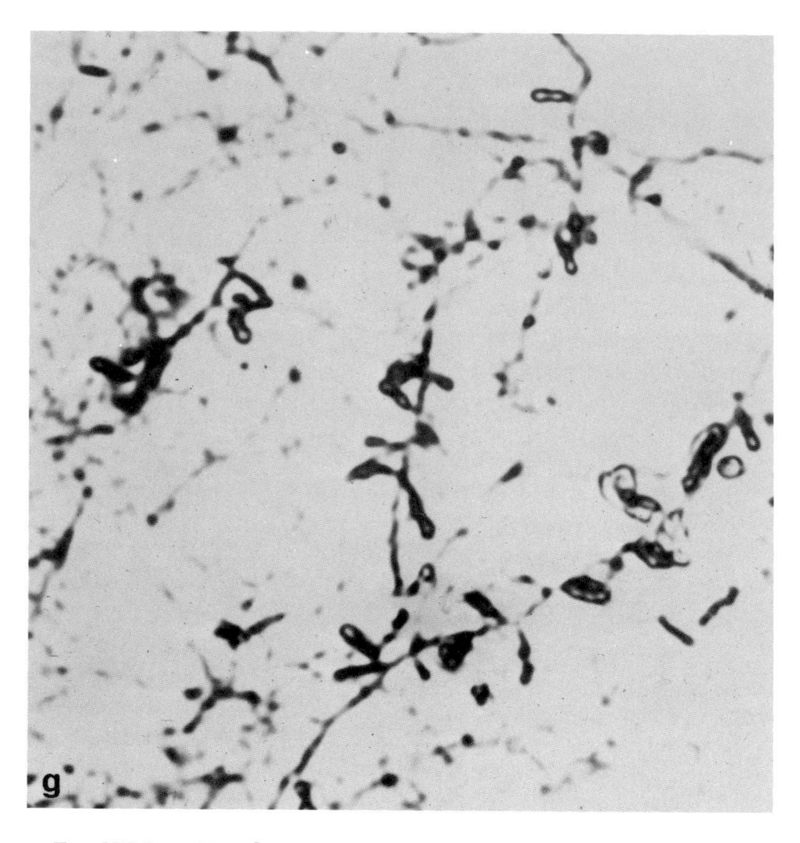

FIG. VII-8g. *Microbispora rosea*, spores on aerial mycelium (x2000).

2. Incubate 5-7 days at 25°C.
3. Acid soils contain more fungi and antifungal agents may be added to the agar to suppress their growth. Glycerol medium also tends to inhibit fungi.

Observation

Look for mycelium and branching filaments in young cultures. Cultures may or may not produce conidia and filaments may or may not break up into rods and cocci in older cultures. [Reference: E. Kuster, and S. T. Williams, *Nature* **202**, 928-29 (1964).]

87. STIMULATION OF BACTERIA AND ACTINOMYCETES BY THE INHIBITION OF FUNGI*

*This procedure may be used with antibacterial antibiotics to bring out the heterotrophic fungi.

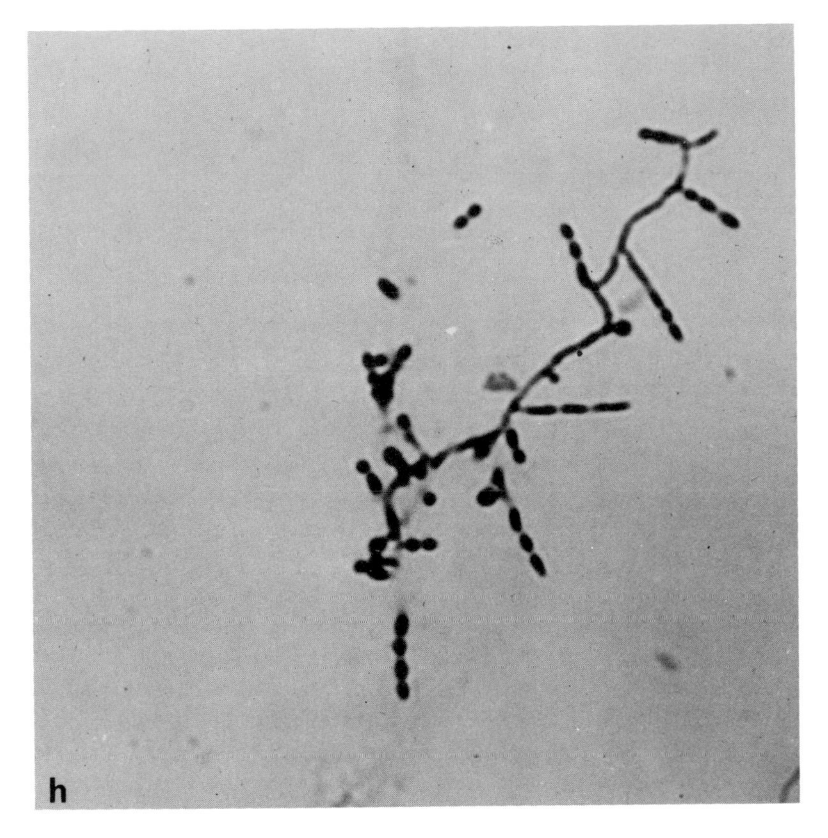

FIG. VII-8h. *Micropolyspora* sp., spores on substrate mycelium (x2000).

Materials

Mycostatin (nystatin), dissolved in dimethyl sulfoxide.
Pimaricin, dissolved in water
Cornmeal agar
Weak peptone-glucose agar

Peptone	0.05 g
Glucose	0.15 g
KH_2PO_4	0.05 g
$MgSO_4 \cdot 7\ H_2O$	0.02 g
$Fe_2(SO_4)_3 \cdot 9\ H_2O$	trace
Agar	1.7 g
Water to	100 ml

Procedure

1. Add antibiotics (pimaricin and/or mycostatin) in freshly prepared

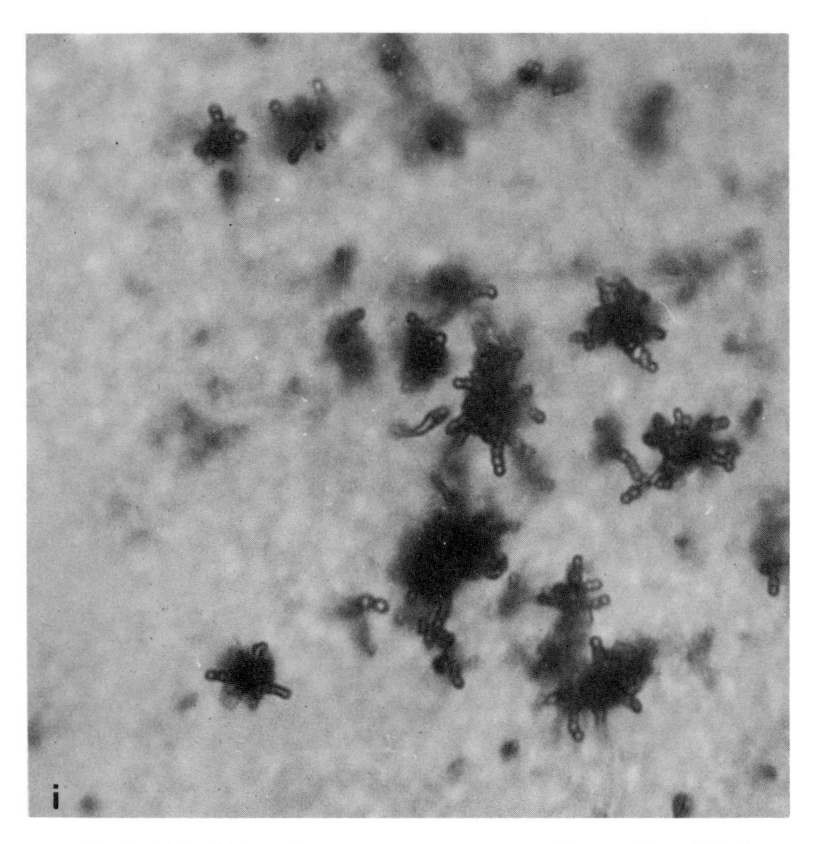

FIG. VII-8i. *Micropolyspora* sp., spores on aerial mycelium (x800).

(filter sterilized is optional) solutions to liquid but cool agar just before pouring Petri dishes.

2. Use pimaricin at a final concentration of 1.7% and mycostatin at a final concentration of 0.5%.
3. After agar is hard, streak agar with soil or water sample.
4. Incubate at 25°C for 4–7 days.

Observation

Presence of antifungal antibiotics will suppress growth of fungi and bring out the various heterotrophic bacteria and actinomycetes present. Colonies should be examined microscopically and cultures isolated and examined biochemically for proper identification. (See

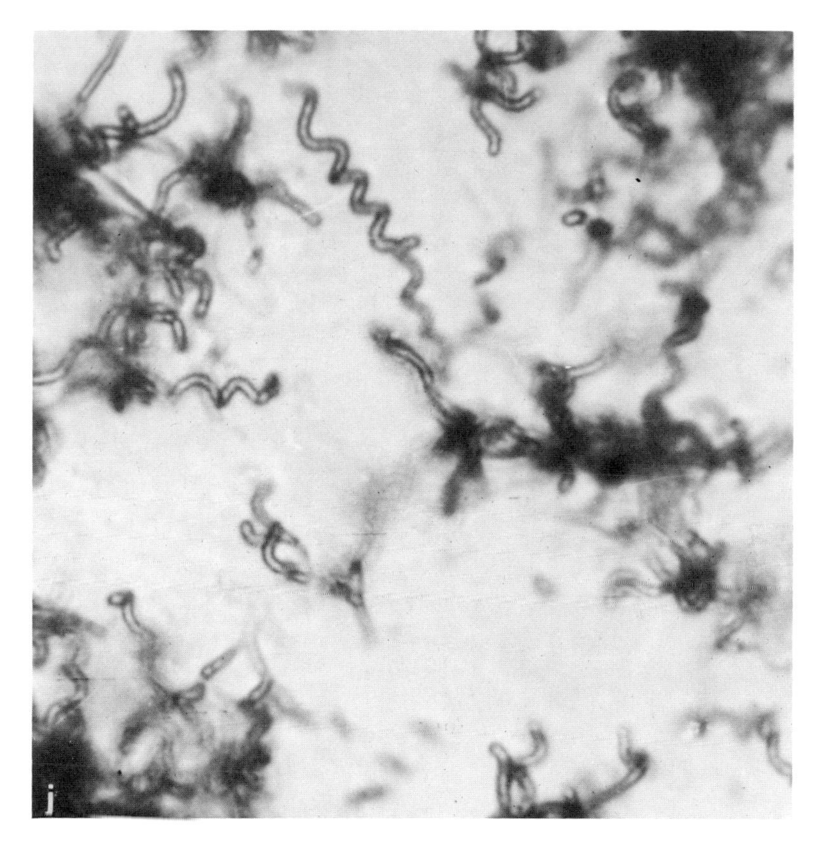

FIG. VII-8j. *Streptomyces* sp., spiral spore chain on aerial mycelium (x900).

Fig. VII-8). [Reference: P. H. Tsao, and D. W. Thieleke, *Can. J. Microbiol.* **12**, 1091-94 (1966).]

M. Slime Bacteria

88. ENRICHMENT AND ISOLATION OF SLIME BACTERIA (MYXOBACTERIA)

Materials

Sterile rabbit dung pellets (or dead bacterial streaks on agar)
Soil or rotting bark
Growth medium

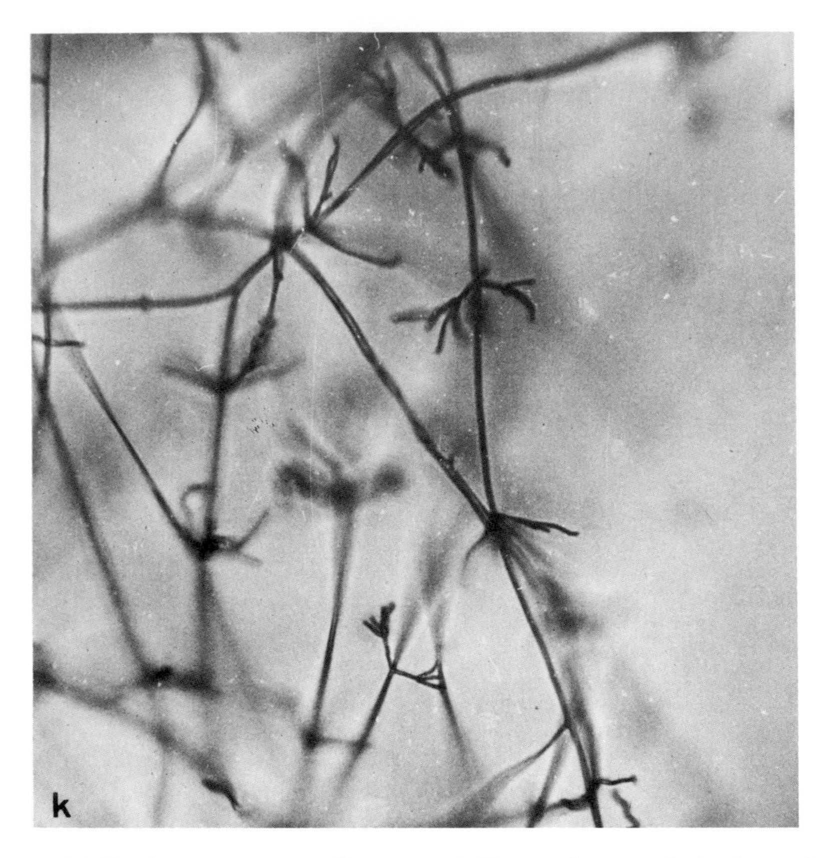

FIG. VII-8k. *Streptomyces* sp. (*Streptoverticillium* sp.), spore chains in verticils on aerial mycelium (x500).

Washed *Escherichia coli* cells	100 ml
$MgSO_4 \cdot 7\ H_2O$	0.05 g
NaCl	0.6 g
Agar	1.5 g
Water to	100 ml
250-ml flask	

Procedure

1. Inoculate sterile rabbit dung pellets (or dead bacteria streaked on agar) with soil sample or rotting bark in a Petri dish.
2. Keep Petri dish in a moist chamber (wet blotter in closed finger-bowl, dessicator) or seal with plastic bag or parafilm to prevent drying.
3. Incubate in the dark at 25°C for 1–4 weeks.

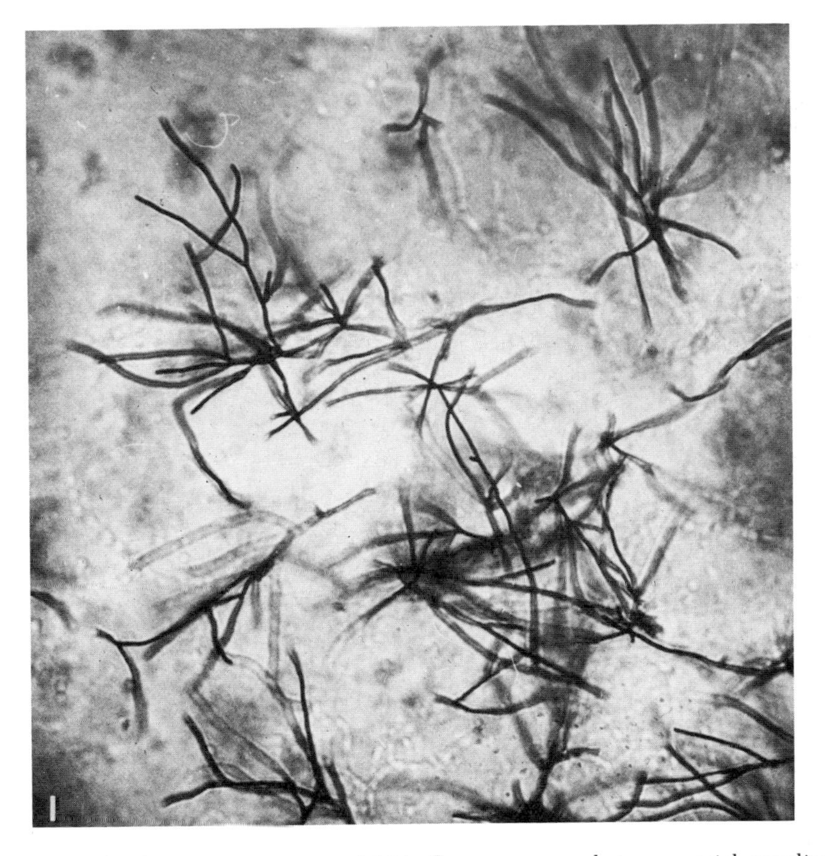

FIG. VII-81. *Streptomyces* sp., straight to flexuous spore chains on aerial mycelium (x500).

Observation

Look for colored fruiting bodies growing as tiny stalks from growth medium under a binocular dissecting microscope (60 ×). Isolate colonies of slime bacteria with a sterile needle and transfer to growth medium plates. Culture of slime bacteria form thin surface colonies with indefinite edges. They may also grow as tiny lens-shaped sub-surface colonies. [Reference: H. D. McCurdy, Jr., *Can. J. Microbiol.* 9, 282–85 (1963).]

89. ENRICHMENT AND ISOLATION OF FRESHWATER AND TERRESTRIAL MYXOBACTERIA

Materials

Enrichment medium

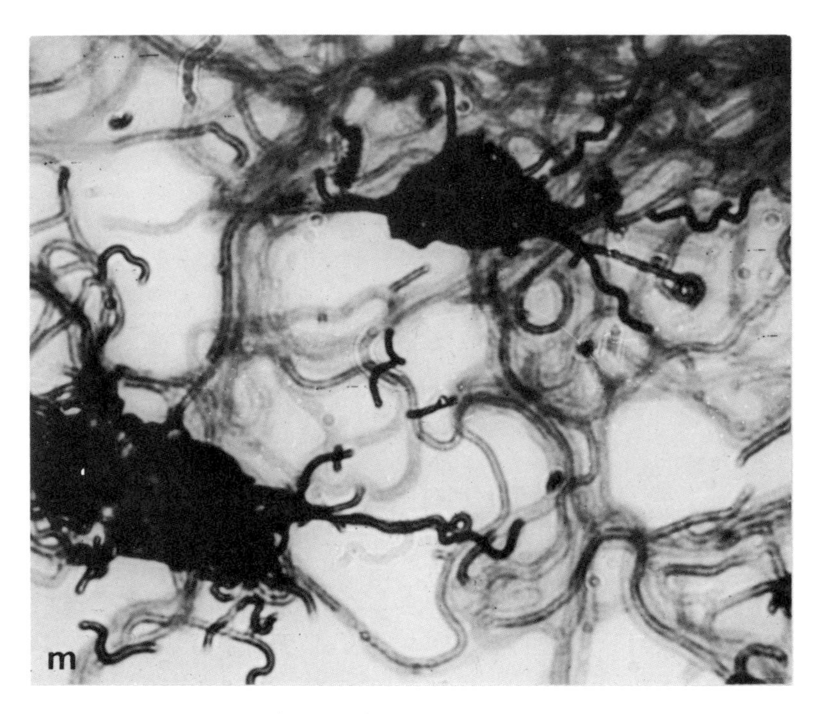

FIG. VII-8m. *Actinopycnidium* sp. (*Streptomyces* sp.), "pycnidia" and spore chains (x750).

(*a*) Rabbit dung pellets or
(*b*) Bacterial streaks on agar medium in Petri dish.

Isolation medium

(*a*) ECM

Washed 24-hour-old *Escherichia coli* cells	100 mg
$MgSO_4 \cdot 7$ H2O	0.05 g
NaCl	0.6 g
Agar	1.5 g
Water to	100 ml

(*b*) SP

Raffinose	0.1 g
Sucrose	0.1 g
Galactose	0.1 g
Soluble starch	0.5 g
Casitone (or casein hydrolyzate)	0.25 g
$MgSO_4 \cdot 7\ H_2O$	0.05 g
K_2HPO_4	0.025 g
Agar	1.5 g
Water to	100 ml

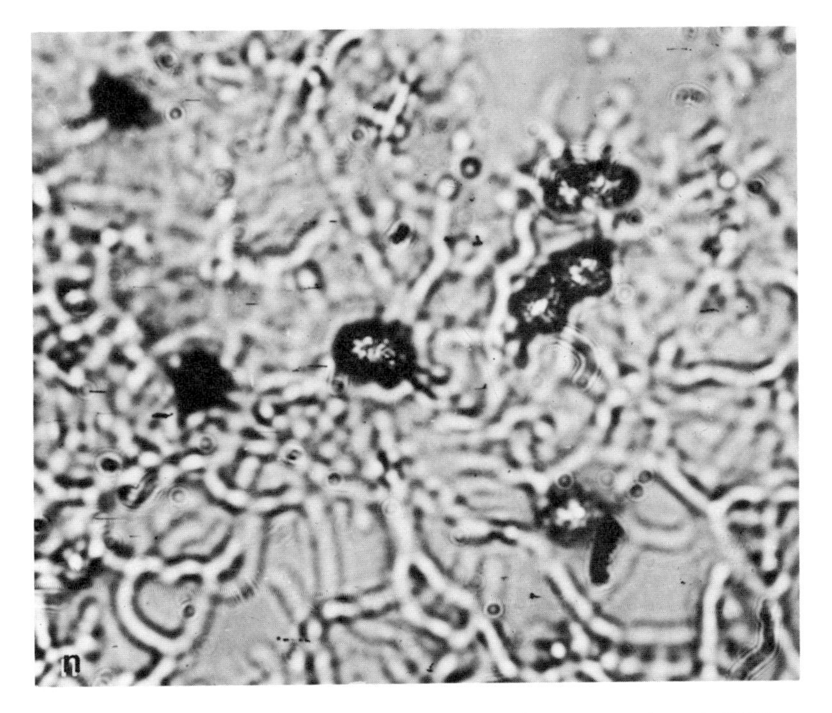

FIG. VII-8n. *Actinoplanes* sp., sporangia on substrate mycelium (x1000).

Vitamin mixture (See Tables VI-9, VI-10, and VI-11.)	0.25 ml
pH	7.4

Soil or bark

Procedure

1. Enrichment for myxobacteria may be made by incubating soil or bark in moist chambers on dung pellets or bacterial streaks on agar medium until fruiting bodies appear.
2. Transfer fruiting bodies to ECM agar plates until pseudoplasmodia extend to a diameter of 3–4 cm.
3. Cut a strip 0.5 × 3 cm from the circumference of the pseudoplasmodium, include the underlying agar.
4. Transfer strip to a sterile mechanical mixer, i.e., Omnimixer, Waring blendor, containing 1.5-g glass beads (for homogenization) and 7.0 ml of a dilution medium containing $MgSO_4 \cdot 7\,H_2O$, 0.05%, K_2HPO_4, 0.025%, and soluble starch, 0.5%.
5. Disperse cells for 90 seconds while vessel is immersed in salt-ice bath at −4°C.
6. Serial dilutions of the dispersed cell suspension in sterile dilution

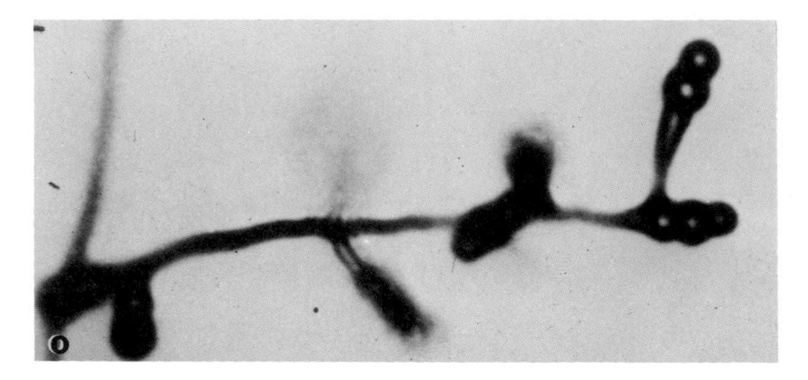

FIG. VII-8o. *Microellobosporia cinera*, sporangia on aerial mycelium (x2000).

medium may be made and aliquots streaked onto EMC or SP medium and incubated at 27°C for 5–10 days.

Observation

The yellow to pink myxobacterial colonies may be identified under binocular dissecting microscope as thin spreading colonies with indefinite edges. Cultures may be purified by restreaking on sterile ECM or SP medium and tested for purity by repeated streaking on brain-heart infusion, trypticase, or tetrathionate agars which do not support the growth of myxobacteria. [Reference: H. D. McCurdy, Jr., *Can. J. Microbiol.* 9, 282–5 (1963).]

90. ISOLATION OF MYXOBACTERIA

Materials

Nutrient media

(*a*) *Peptonized milk*	0.1 g
Agar	1.5 g
Water to	100 ml
(*b*) *Nutrient medium A*	100 ml

Sterilize and add each of the following from sterile solutions: (use a minimal volume)

$(NH_4)_2SO_4$	0.01 g
$MgSO_4 \cdot 7\ H_2O$	0.05 g

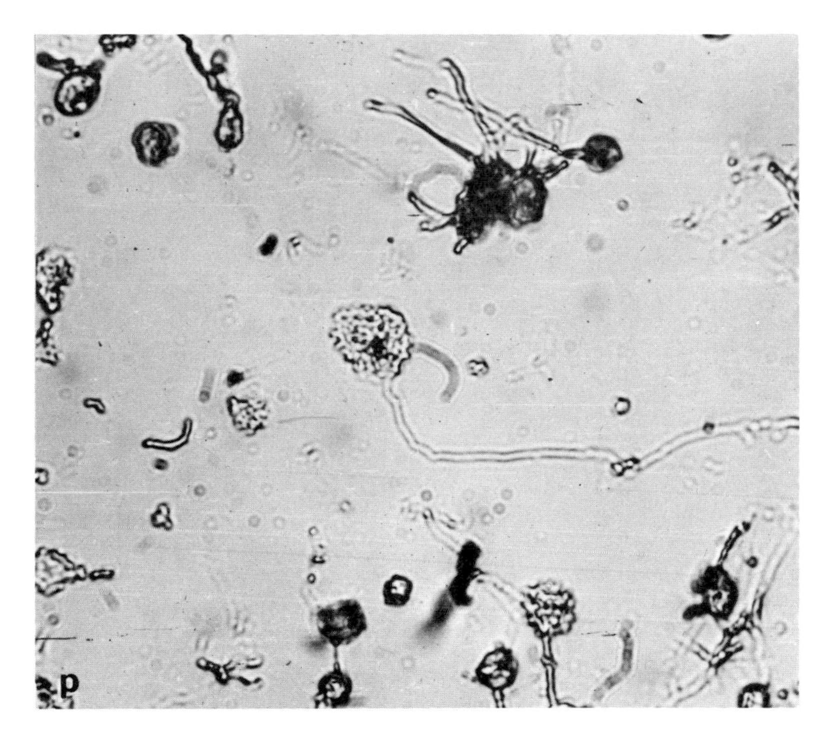

FIG. VII-8p. *Streptosporangium* sp., sporangia on aerial mycelium (x1000).

$FeCl_3 \cdot 6\ H_2O$	0.001 g
$CaCl_2$	0.025 g
$MnCl_2$	0.01 mg
K_2HPO_4	0.025 g
(*c*) *Peptonized milk*	0.1 g
Agar	1.5 g
Crystal violet (dissolve in alcohol)	1.0 mg
Water to	100 ml

Procedure

1. Sterilize any one of the above media and pour into sterile Petri dishes.
2. Streak surface of agar with material from water or soil sample.
3. Incubate in dark for 1 week.
4. Medium (a) is best for most myxobacteria; medium (b) is best for maintenance; medium (c) inhibits growth of motile eubacteria.

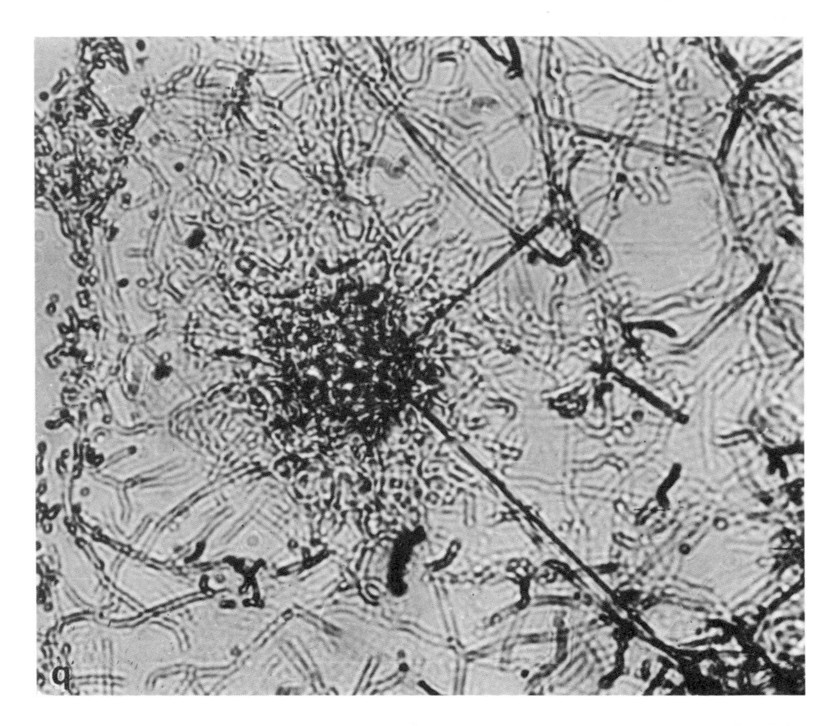

FIG. VII-8q. *Actinosporangium* sp. (*Streptomyces* sp.), sporangium (x450).

Observation

Use binocular dissecting microscope to look for thin surface colonies with indefinite edges. Colonies often produce colored fruiting bodies as tiny stalks. Transfer bacteria or fruiting body to fresh media to obtain pure cultures. [Reference: E. E. Jeffer, *Intern. Bull. Bacteriol. Nomenclature Taxonomy* **14**, 115-36 (1964).]

91. THE ISOLATION OF THE MYXOBACTER, *Chondromyces*

Materials

Decaying wood
Oatmeal

Boil 25 g of oatmeal in 500 ml water for 10 minutes. Filter through cheese cloth. Bring volume of filtrate to 350 ml with water.

Nutrient agar medium

Oatmeal filtrate	0.2 ml
Peptonized milk	0.02 g

Agar	1.0 g
Water to	100 ml
pH	7.4

Procedure

1. Place decaying wood in a moist chamber and incubate at room temperature (20°C) for 4-7 days.
2. Examine for fruiting bodies.
3. Transfer fruiting bodies with sterile rod or loop to nutrient agar medium and incubate as before.

Observation

Examine agar medium for presence of thin translucent colonies with indeterminate edges. Myxobacteria may be purified by isolating a group of bacteria advancing before slime front by cutting agar between them and slime. The isolated bacteria may be transferred to sterile media and tested for purity by repeated restreaking of culture material on media for the growth of heterotrophic bacteria. [Reference: L. F. Nellis, and H. R. Garner, *J. Bacteriol.* **87**, 230-31 (1964).]

92. Isolation of Myxobacteria

Materials

Medium

Tryptone	0.05 g
Yeast extract	0.05 g
Beef extract	0.02 g
Sodium acetate · 3 H_2O	0.02 g
Agar	0.9 g
Water (or seawater) to	100 ml
pH	7.2-7.4

Procedure

1. Plate 0.1-ml water sample on surface of Petri plates containing above agar medium.
2. Incubate for 3 days at room temperature.

Observation

Cytophaga colonies on this medium are translucent, irridescent, and contain patterns of light and dark areas due to the diffraction of light. *Chondrococcus* form spreading, rhizoid, greenish-yellow colonies which often have a honeycomb appearance. [Reference: R. L. Anderson, and E. J. Ordal, *J. Bacteriol.* **81**, 130–138 (1961).]

93. Isolation of Anaerobic Marine Agar-Decomposing Myxobacteria

Materials

Nutrient medium (may be solidified with agar, 2 g/100 ml)	
$NaCl$	3.0 g
KH_2PO_4	0.1 g
NH_4Cl	0.1 g
$MgCl_2 \cdot 6\,H_2O$	0.05 g
$CaCl_2$	0.004 g
$NaHCO_3$	0.5 g
$Na_2S \cdot 9\,H_2O$	0.01 g
Fe citrate (0.004 *M*)	0.5 ml
Trace elements*	0.2 ml
Agar	0.5 g
Yeast extract	0.03 g
Water to	100 ml
pH	7.0
Aerobic medium	
$NaHCO_3$	0.05 g
Nutrient broth	0.1 g
Corn steep liquor	0.1 g
Yeast extract	0.1 g
Agar	2.0 g
Water (or seawater) to	100 ml
pH	7.0

Procedure

1. Inoculate bottles completely filled with nutrient medium with marine mud taken from area where algae were decaying.
2. Stopper bottles and incubate in the dark at 30°C for 3–5 days.
3. Inoculate material from bottles into anaerobic plates or shake cultures containing nutrient medium plus 2% agar.
4. Cover shake cultures with sterile paraffin to maintain anaerobiosis.

*Trace elements: H_3BO_3, 0.28 g; $MnSO_4 \cdot 6\,H_2O$, 0.21 g; $Cu(NO_3)_2 \cdot 3\,H_2O$, 0.02 g; $Na_2MoO_4 \cdot 2\,H_2O$, 0.075 g; $CoCl_2 \cdot 6\,H_2O$, 0.02 g; $Zn(NO_3)_2 \cdot 6\,H_2O$, 0.025 g; Water to 100 ml.

Observation

Examine colonies of myxobacteria microscopically (100–400 X) for flexuous movements of myxobacteria. These colonies may be purified by further streaking on anoerobic agar or inoculation into shake cultures. Some myxobacteria grow well anaerobically only if carbonate is supplied.

Agar digestion may be seen anaerobically on nutrient medium or aerobically on aerobic medium if 1% agar is used to solidify the medium. Colonies may appear in depressions in agar or if agar is flooded with a solution of I + KI, colonies will be surrounded by clear gelase fields. [Reference: H. Veldkamp, *J. Gen. Microbiol.* **26**, 331–342 (1961).]

N. Spirochetes

94. Isolation of Leptospiras (Spirochetes)

Materials

Nutrient media

(*a*) Human feces	1.0 g
Water	20.0 ml
(*b*) Egg yolk (raw)	1.0 g
Agar	1.0 g
Water to	300 ml
(*c*) Nutrient agar	0.3 g
Rabbit serum (add aseptically)	10.0 ml
Water to	100 ml
(*d*) Zuelger's medium*	
Agar	1.0 g
Brilliant green	3.0 mg
KH_2PO_4	183 mg
$Na_2HPO_4 \cdot 7\ H_2O$	650 mg
Water to	300 ml
pH	7.5

Procedure

1. Water samples may be inoculated into any one of the above media for the cultivation of spirochetes. Advantage may be taken of their ability to move through dilute agar media and thus move away

*After sterilization add 1.0 ml sterile egg yolk.

from their contaminants. Spirochetes are very thin and flexuous which permit them to pass through filters (0.2 μ, diameter of pore size) which prevent ordinary bacteria from passing.
2. This filtrate may be inoculated into any one of the above sterile media.

Observation

Spirochetes present in inoculum penetrate agar and form spreading colonies beyond edge of inoculum. They become visible after several weeks incubation at 20°-25°C. Cultures may be purified by the same procedure—filtration and inoculation into above media. Some of these media will also permit the growth of pathogenic leptospiras although the saprophytic forms usually outgrow the pathogens. [Reference: H. Veldkamp, 1965. *Zentralbl. Bakteriol. Parasitenk. Suppl.* 1, 407-14 (1965).]

95. Isolation of Marine Anaerobic Spirochetes

Materials

Nutrient agar medium	
$(NH_4)_2SO_4$	0.1 g
$MgSO_4 \cdot 7\,H_2O$	0.05 g
K_2HPO_4	0.01 g
Yeast extract	0.01 g
Agar	0.3 g
Seawater to	100 ml
Agar-shake culture	
Nutrient medium above to	100 ml
Agar	1.0 g
$Na_2S \cdot 9\,H_2O$	0.01 g

Procedure

1. Mix a small sample of marine mud with 30 ml of nutrient agar medium in a 100-ml Erlenmeyer flask just before agar hardens.
2. Incubate at room temperature for 3 weeks.

Observation

Upper layer of agar appears softened through decomposition by aerobic rods and spirilla. Below this surface there may appear spherical colonies of low density formed by spirochetes. These may be subcultured by removing the upper aerobic layer with suction and inoculating spirochete material into agar shake cultures. These may be kept anaerobic by overlaying the tubes with sterile paraffin or by other anaerobic methods. [Reference: H. Veldkamp, *Zentralbl. Bakteriol. Parasitenk. Suppl.* 1, 407-414 (1965).]

O. Miscellaneous Bacteria

96. Enrichment and Isolation of *Leucothrix*

Materials

Ulva (or other marine algae)	
Nutrient medium (after Harold and Stanier) (agar 2 g/100 ml for solid media)	
Tryptone	0.04 g
Yeast extract	0.04 g
Beef extract	0.02 g
Sodium acetate · 3 H_2O	0.02 g
Synthetic seawater (or seawater) to	100 ml
pH	8.0-8.3
Defined medium (after Brock)	
NaCl	1.175 g
$MgCl_2 \cdot 6\ H_2O$	0.535 g
Na_2SO_4	0.2 g
$CaCl_2 \cdot 2\ H_2O$	0.075 g
KCl	0.035 g
Tris(hydroxymethyl)aminomethane	0.05 g
Na_2HPO_4	0.005 g
Na H glutamate	0.1 g
Agar	1.5 g
Water to	100 ml
pH	7.6

Procedure

1. Place pieces of *Ulva* or other marine algae in a flask.
2. Fill the flask almost to the top with seawater.
3. Incubate at 25°C in diffuse daylight for 8-12 days or until a pellicle is formed.

Observation

Examine pellicle microscopically for long, colorless, unbranched, tapering threads. Threads may form a rosette. Pure cultures may be isolated by streaking material from the pellicle or from the surface of marine algae on to agar plates containing either of above media and incubating at 25°C for 1–2 weeks. Examine surface of plate during incubation period with a binocular dissecting microscope (125 ×) for characteristic "finger-print" colonies. Transfer "finger-print" colonies to fresh media until contaminating colonies are no longer seen on solid media. [Reference: (1) R. Harold, and R. Y. Stanier, *Bacteriol. Rev.* **19**, 49–64 (1955); (2) T. D. Brock, *Limnol. Oceanogr.* **11**, 303–307 (1966).]

97. Isolation of Sheath-Forming Bacterium, *Sphaerotilus*

Materials

Slime masses from polluted stream, lakes, or activated sludge

(*a*) *Isolation medium* (*CGY*) (agar, 1.5 g/100 ml may be added to harden medium)

Pancreatic digest of casein (Trypticase or Casitone)	0.5 g
Glycerol	1.0 g
Yeast autolyzate	0.1 g
Water to	100 ml

(*b*) *Isolation medium* (*GG*) (agar 1.5 g/100 ml may be added to harden medium)

Glycerol	0.5 g
L-Glutamic acid	0.09 g
$MgSO_4 \cdot 7\,H_2O$	0.01 g
$FeSO_4 \cdot 7\,H_2O$	0.05 g
$CaCl_2 \cdot 2\,H_2O$	3.0 mg
$ZnSO_4 \cdot 7\,H_2O$	3.0 mg
Water to	90 ml
pH (adjusted with KOH)	7.0

Sterilize phosphate solution separately (K_2HPO_4, 1.14 g, KH_2PO_4, 0.46 g, water to 100 ml) and add 10 ml to sterile medium GG.

Procedure

1. Slime masses are washed well in running water and teased into small portions.

2. These are homogenized in a blender for 30–45 seconds in a minimum of suspending liquid.
3. The homogenate is streaked on agar plates of the isolation medium (CGY or GG) and incubated for 24–48 hours at 28°C (48 hours or more at 20°C).

Observation

Examine colonies under microscope (100 ×) for tangled curled filamentous growth. To confirm, examine smears or wet mounts microscopically at 900–1000 × for filaments of cylindrical sheaths enclosing ellipsoidal or rod-shaped cells; cells lack chlorophyll. Reisolation of colonies may be used to insure purity of cultures. (See Fig. VII-9.) [Reference: N. C. Dondero, R. A. Phillips, and H. Heukelekian, *Appl. Microbiol.* **9**, 219–27 (1961).]

98. Isolation of the Floc-Forming Bacterium, *Zoogloea*

Materials

Isolation medium	
Protease-peptone	0.2 g
Yeast extract	0.1 g
Water to	100 ml
pH	7.0

Procedure

1. Raw water sample, usually from sewage, is inoculated into sterile isolation medium and the medium is aerated with sterile moist air until flocs appear. Flocs (or zoogleal masses) are bacteria embedded in a slimy matrix.
2. Single flocs are separated and suspended in sterile water by shaking with sterile glass beads. This suspension is diluted.
3. One-milliliter aliquots of 10^{-6}, 10^{-7}, 10^{-8} dilutions are added to sterile isolation medium in tubes and aerated at 28°C for 48 hours with moist sterile air.
4. Tubes showing turbidity before 48 hours are discarded. Tubes containing only flocculent growth are carried through this dilution procedure again or until culture purity is certain.

Observation

Zoogloea form finger-like projections in a wet mount. Cells are found in a slimy matrix; they are gram-negative rods with rounded ends 1–2 × 4 μ in size. Occasional rods are motile with a single polar

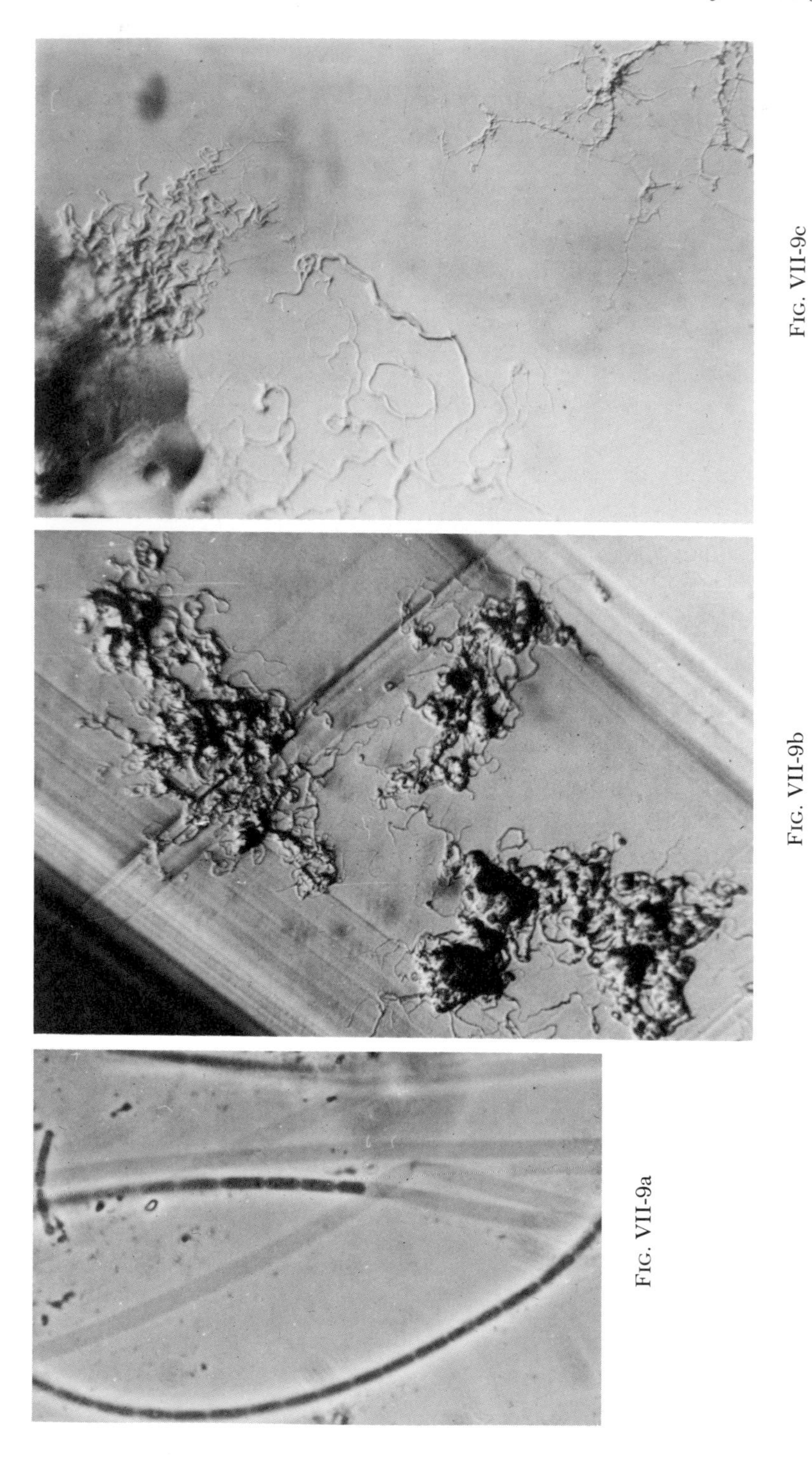

FIG. VII-9c

FIG. VII-9b

FIG. VII-9a

Fig. VII-9d

Fig. VII-9. Appearance of *Sphaerotilus*. [From Dondero, N. C., Phillips, R. A., and Heukelian, H., *Appl. Microbiol.* **9**, 219–227 (1961).] (a). Sheaths and cells within the sheaths; phase contrast. (b). Young *Sphaerotilus* colonies; oblique transmitted illumination. (c). Typical *Sphaerotilus* growth; oblique transmitted illumination. (d). Method of inoculating agar for isolation of *Sphaerotilus*. Examine edges of the streak.

flagellum (See Fig. VII-10.) [Reference: P. R. Dugan, and D. G. Lundgren, *Appl. Microbiol.* **8**, 357–361 (1960).]

99. Isolation of *Bacteroides* and Related Bacteria

Materials

Nutrient medium

Sodium azide	0.009 g
Sodium deoxycholate	0.07 g
Ethyl violet	0.0007 g

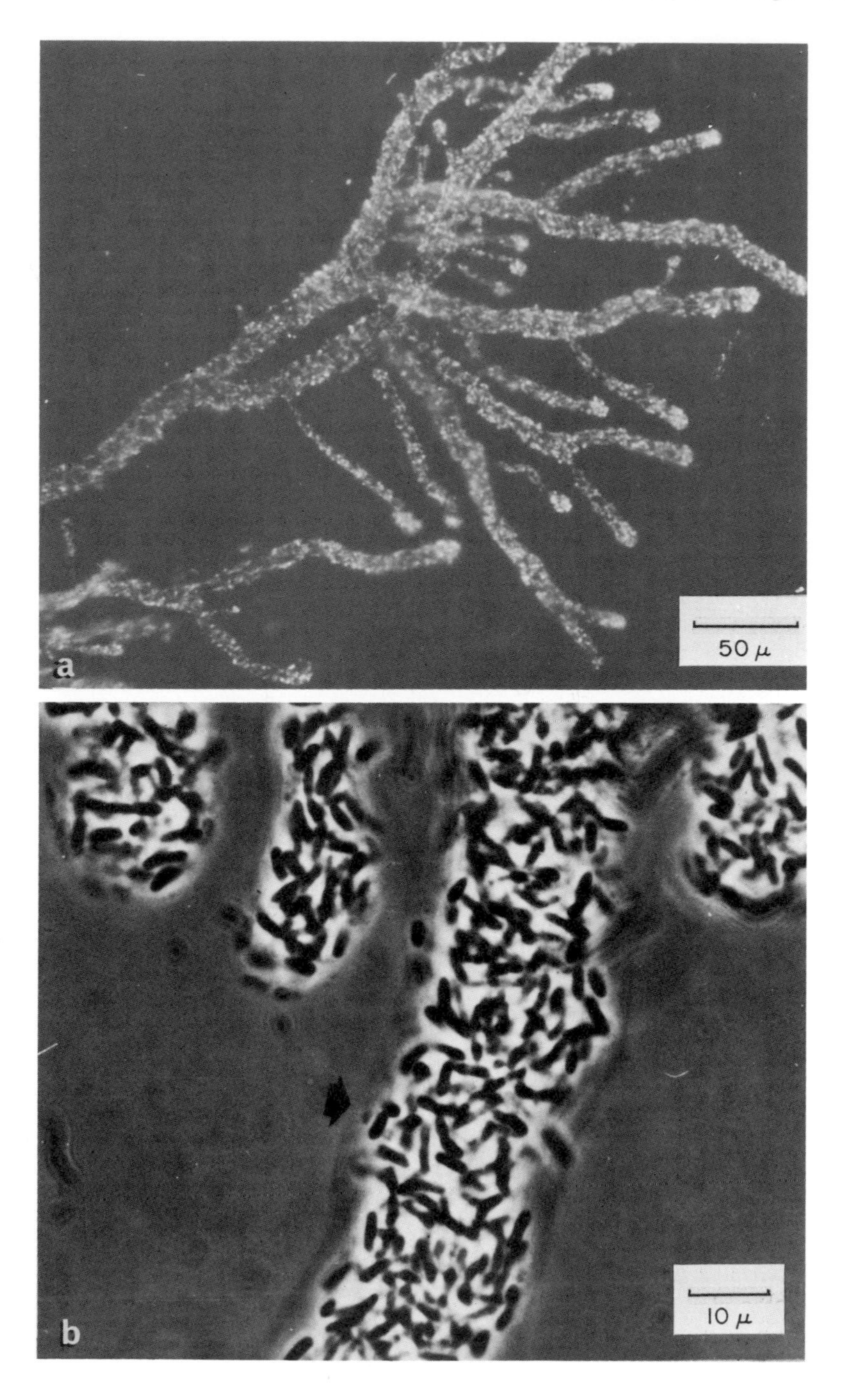
50 μ
a
10 μ
b

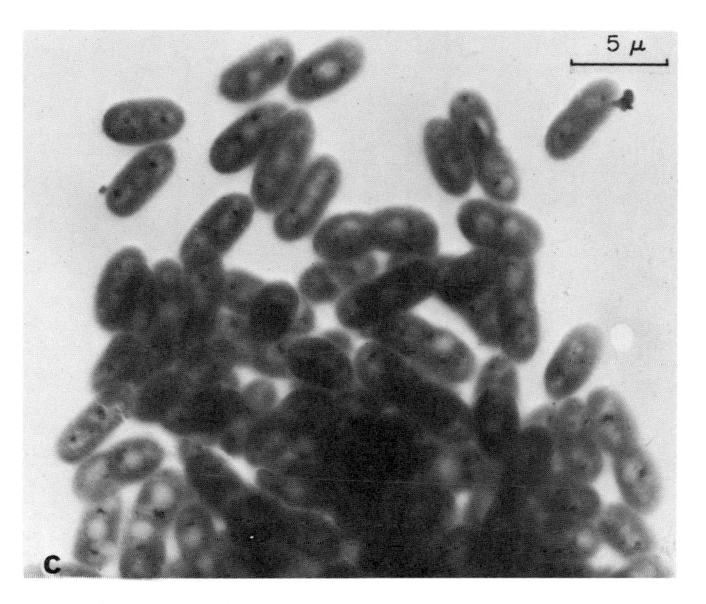

FIG. VII-10. Illustrations of *Zoogloea*. [From Unz, R. F., and Dondero, N. C., *Can. J. Botany* 13, 1671–1682 (1967).] (a). Typical *Zoogloea*, dark field, wet mount. (b). Cells embedded in matrix (seen as pale margin at arrow); dark field, wet mount. (c). Tip of zoogloeal branch showing individual cells. Electron micrograph.

Brain heart infusion agar (Difco) to 100 ml
Anaerobic jars
Nitrogen and CO_2
Anaerobiosis indicator

Ten milliliters of a freshly boiled solution of the following in a tube or beaker: 6 ml of 0.1 *N* NaOH diluted to 100 ml, glucose, 6 g, and methylene blue, 0.015 g.

Procedure

1. Minimize exposure of sample to air.
2. Dilutions of water, sewage, etc., are prepared and 0.1-ml portions are streaked on the surface of nutrient agar with sterile L-shaped glass rod.
3. Place plates in anaerobic jars.
4. Evacuate to 700 mm. Hg and flush jars twice with nitrogen and then evacuate and replace with 90% N_2 and 10% CO_2.

Incubate at 37°C for 4–5 days.
Anaeriobiosis may be visualized by adding methylene blue indicator.

Observation

Colonies may be blue or white, 1-2 to 3-5 mm in diameter. Microscopically the *Bacteroides* group are gram-negative rods with rounded ends, obligately anaerobic, and nonspore formers. Pleomorphic cells with globular forms, branching with filaments, 0.5-1 μ wide by 1-3 μ long, that stain unevenly and bipolarly are *Sphaerophorus*. Nonpleomorphic cells occur singly, in pairs, and occasionally in short chains, 1-2 μ long by 0.5-1 μ wide; cells staining evenly or bipolarly are *Bacteroides*. [Reference: F. J. Post, A. D. Allen, and T. C. Reid, *Appl. Microbiol.* **15**, 213–18 (1967).]

100. Isolation of *Bdellovibrio*, a Predatory, Parasitic Bacterium

Materials

Millipore filter
Millipore filter membranes with pore sizes: 3, 1.2, 0.8, 0.65, and 0.45 μ
YP medium

Yeast extract	0.3 g
Peptone	0.06 g
Water to	100 ml
pH	7.2

Any one or all of the following indicator bacteria*
Erwinia amylovora EA137
Escherichia coli B 2262
Aerobacter aerogenes 2001
Proteus mirabilis
Serratia marcescens 2031
Pseudomonas fluorescens ATCC 12633

Procedure

1. Suspend a 500-g soil sample in 500 ml of water and shake vigorously for 1 hours.
2. Centrifuge suspension for 5 minutes at 2000 rpm.
3. The dirty supernatant is filtered stepwise starting with the largest

*See reference for other useful bacteria

pore size 3 μ and proceeding through 1.2, 0.8, 0.65, and 0.45 μ pore size in order. Portions of the last two fractions which have relatively few bacteria are then mixed with a concentrated 24-hour-old suspension (in water) of the indicator bacterium by the double-layer technique used for the isolation of bacteriophage. See below.

4. Plates are prepared by pouring a bottom layer of YP medium containing 1.9% agar. Allow this to harden.
5. Portions of the filtered fractions (1.0 ml) and one drop of concentrated indicator bacterial suspension in YP medium are mixed with a small volume of melted, but cool, YP medium containing 0.6% agar (4.0 ml).
6. This suspension is mixed well and poured on the bottom layer of YP agar to make a second softer layer. The Petri dishes must be horizontal while the agar hardens to permit a confluent upper layer.
7. Petri dishes are incubated at 20–28°C.
8. Plaques (clear areas due to bacterial lysis) observed after 24 hours are probably due to bacteriophages.
9. These are marked and Petri dishes are incubated for 2–4 days. Plaques forming after 2 days may contain *Bdellovibrio.*

Observation

Suspected *Bdellovibrio* plaques (Fig. VII-11) are lifted out and suspended in liquid YP medium and examined by phase microscopy (1000 ×). The parasitic bacterium is a tiny gram-negative curved rod with a thick polar flagellum. They are about 0.3 μ in width and of variable length. They may appear attached to their host bacteria. Strains may be purified by carrying them through the isolation procedure described above three times. [Reference: H. Stolp, and M. P. Starr, *Antonie van Leeuwenhoek J. Microbiol. Serol.* **29**, 217–248 (1963).]

P. Isolation of Bacteria which Excrete Metabolites

101. Isolation of Microorganisms which Excrete Metabolites by use of Antimetabolites in Gradient Plates

Materials

Antimetabolite (as concentrated as possible) in 1.5% agar (See Table III-5.)

Nutrient agar medium for various microorganisms, i.e., algae, fungi, bacteria

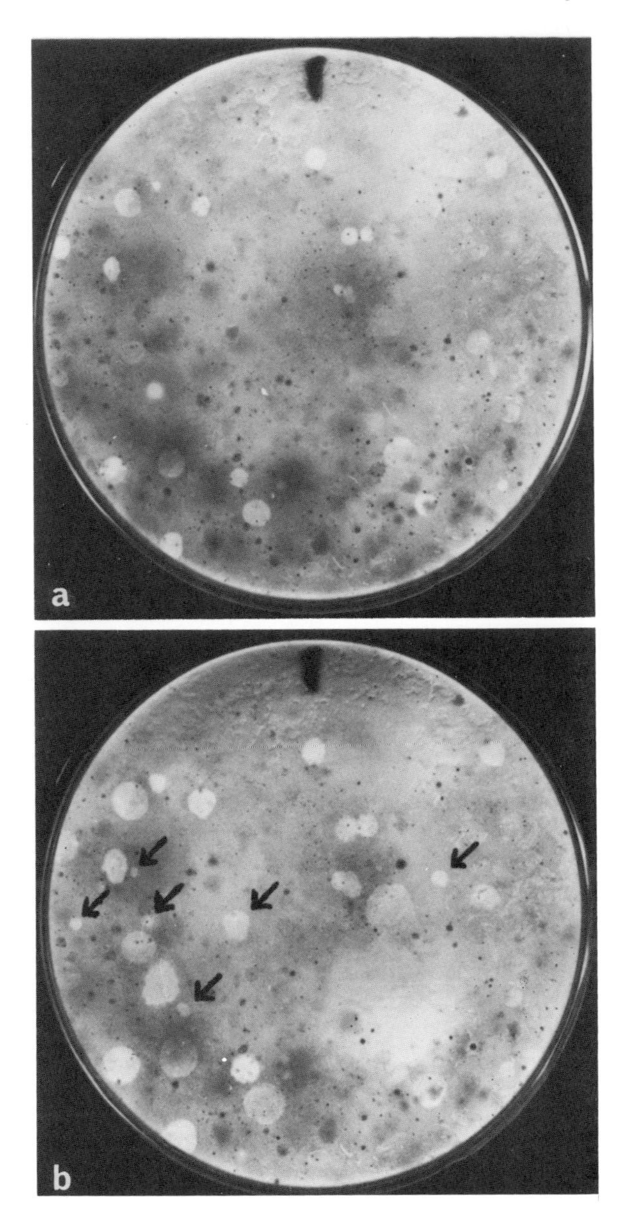

FIG. VII-11. Plaques of *Bdellovibrio bacteriovirus*. [From Klein, D. A., and Casida, L. E., Jr., *Can. J. Microbiol.* **13**, 1235–1241 (1961).] (a). Plaques after 4 days of incubation on *E. coli* double layer plate. (b). Plaques after 6 days of incubation on *E. coli* double layer plate. Arrows indicate plaques present at 6 but not 4 days of incubation.

Procedure

1. Prepare 10-ml aliquots of concentrated solution of antimetabolite with 1.5% agar.
2. Sterilize and pour into sterile Petri dish.
3. Harden agar so that agar forms a slope by raising one end of Petri dish (see Fig. VII-12). Mark end with low end of slope on outside of bottom dish.
4. Pour 10 ml of sterile melted selected agar medium into Petri dish held horizontally so as to form a flat agar field.
5. Harden plates and streak them with samples of soil or water.
6. Incubate under conditions appropriate to isolate desired microorganisms.

Observation

Look for colonies of microbes capable of overcoming inhibition due to presence of antimetabolite by secreting excess metabolite. [Reference: Modified from G. H. Scherr, and M. E. Rafelson, Jr. *J. Appl. Bacteriol.* **25**, 187–194 (1962).]

102. Isolation of Microorganisms Which Excrete Metabolites by Use of Antimetabolites

Materials

Nutrient agar medium preferably chemically defined
Antimetabolite (see antimetabolites, Table III-5)

Procedure

1. Inoculate soil or water sample into liquid media containing increasing concentrations of antimetabolite for enrichment.
2. Prepare nutrient agar plates containing 0.5% (concentration varies with compound) of antimetabolite.
3. Streak surface of plate with soil water sample or enrichment.
4. Incubate until colonies appear.

Observation

Colonies of resistant microorganisms growing in presence of high concentrations of antimetabolite may overcome inhibition by synthesizing large quantities of the corresponding metabolite. The production of the metabolite may be confirmed by growing resistant strain in presence of antimetabolite in chemically defined liquid

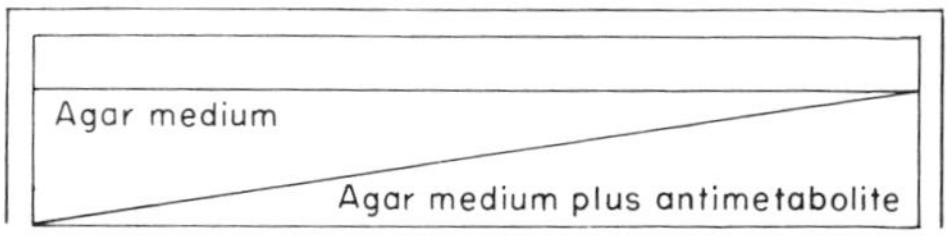

FIG. VII-12. Concentration gradient plate for the isolation of microorganisms capable of growing in the presence of high concentrations of antimetabolite.

medium and testing the medium for the presence of metabolite by thin-layer or paper chromatography or feeding a known auxotroph (mutant strain requiring metabolite). [Reference: Modified from E. A. Adelberg, *J. Bacteriol.* **76**, 326 (1958).]

103. Identification of Microorganisms Which Excrete Metabolites by Bioautography

Materials

Indicator organism. An auxotroph or prototroph which requires a specific metabolite
Number of isolated microorganisms
Nutrient agar medium for indicator organism
Nutrient agar medium for isolates

Procedure

1. Prepare large Petri dish with nutrient agar medium for growth of isolates.
2. Inoculate isolated microorganisms in known positions on nutrient agar.
3. Incubate until colonies are formed.
4. Kill colonies with UV light.
5. Overlay colonies with indicator nutrient agar containing indicator organisms.
6. Incubate or just overlay.

Observation

Indicator organism for specific metabolite will grow over and around colonies producing required metabolite and nowhere else. [Reference: Modified from S. Udaka, *J. Bacteriol.* **79**, 754-5 (1960).]

III. Fungi

104. Isolation Media for Fungi

Materials

Conn's agar	
KNO_3	0.2 g
$MgSO_4 \cdot 7\ H_2O$	0.12 g
KH_2PO_4	0.27 g
Maltose	0.72 g
Potato starch	1.0 g
Agar	1.5 g
Water to	100 ml
Cornmeal agar	
Cornmeal	2–6 g
Peptone (optional)	2.0 g
Glucose (optional)	2.0 g
Agar	1.5 g
Water to	100 ml
Czapek (Dox) agar	
$NaNO_3$	0.2 g
K_2HPO_4	0.1 g
KCl	0.05 g
$MgSO_4 \cdot 7\ H_2O$	0.05 g
$FeSO_4$	1.0 mg
Sucrose	3.0 g
Agar	1.5 g
Water to	100 ml
Malt agar (use glucose for Mucoraceae)	
Malt extract	2.5 g
Agar	1.5 g
Water to	100 ml
Sabouraud's agar	
Peptone	1.0 g
Maltose (or glucose)	4.0 g
Agar	1.5 g
Water to	100 ml

[Reference: G. C. Ainsworth, "Dictionary of the Fungi." Commonwealth Mycological Institute, Kew, Surrey, England, 1961.]

105. ISOLATION OF STARCH-DIGESTING (AMYLASE) BACTERIA AND FUNGI (See Method No. 30)

106. ENRICHMENT AND ISOLATION OF CHITINOLYTIC MICROORGANISMS (BACTERIA OR FUNGI)—Campbell and Williams (See Method No. 21)

107. ENRICHMENT AND ISOLATION OF CELLULOLYTIC MICROORGANISMS (BACTERIA OR FUNGI) (See Method No. 16)

108. ISOLATION OF MARINE PHYCOMYCETES

Materials

Glucose	0.1 g
Gelatin hydrolyzate	0.1 g
Liver extract (1:20)	0.001 g
Yeast extract	0.01 g
Agar	1.2 g
Seawater to	100 ml

Procedure

1. Sterilize medium and while agar is melted, but not too hot, add 0.05 g/100 ml of streptomycin sulfate and 0.05 g/100 ml of Na penicillin G.
2. Mix well and pour agar medium into sterile Petri dishes.
3. Collect marine algae and place small pieces of algae on agar surface.
4. Incubate Petri dishes in the dark at 20°C for 3-14 days.

Observation

Look for filamentous fungi growing away from algal inoculum. Transfer filaments to fresh agar medium. Purity may be tested by growing fungus in above medium minus agar and antibiotics. [Reference: M. S. Fuller, B. E. Fowles, and D. J. McLaughlin. *Mycologia* **56**, 745-56 (1964).]

109. MEDIA FOR THE GROWTH OF MARINE FUNGI

Materials

Dermocystidium medium

KCl	0.07 g
K_2HPO_4	0.043 g

$CaCl_2$	0.047 g
NaCl	2.4 g
$MgSO_4 \cdot 7\ H_2O$	0.8 g
Casein hydrolyzate (enzymatic)	0.2 g
Glucose	0.4 g
Thiamine · HCl	20 μg
Vitamin B_{12}	0.3 μg
Na_2 EDTA	1.0 mg
B (as H_3BO_3)	0.2 mg
Fe (as $FeCl_3 \cdot 6\ H_2O$)	0.19 mg
Mn (as $MnCl_2 \cdot 4\ H_2O$)	0.1 mg
Mo (as $Na_2MoO_4 \cdot 2\ H_2O$)	0.09 mg
Zn (as $ZnCl_2$)	0.06 mg
Co (as $CoCl_2 \cdot 6\ H_2O$)	1.0 mg
Cu (as $CuCl_2 \cdot 2\ H_2O$)	2.0 μg
Agar	0.1 g
Water to	100 ml
pH (after autoclaving)	7.3
Modified ASP6 medium	
K_2HPO_4	15.0 mg
NaCl	2.4 g
$MgSO_4 \cdot 7\ H_2O$	0.8 g
KCl	0.07 g
$CaCl_2$	41.6 mg
Tris(hydroxymethyl)amino-methane	0.2 g
Na H glutamate	0.2 g
Glucose	0.2 g
Na_2 glycerophosphate · 5 H_2O	20.0 mg
Vitamin B_{12}	1.0 μg
Biotin	0.05 mg
Thiamine · HCl	0.02 mg
Nicotinic acid	0.01 mg
EDTA	1.0 mg
$FeCl_3 \cdot 6\ H_2O$	1.0 mg
$ZnCl_2$	0.13 mg
$MnCl_2 \cdot 4\ H_2O$	0.36 mg
$CoCl_2 \cdot 6\ H_2O$	4.0 mg
H_3BO_3	1.14 mg
$NaMoO_4 \cdot 2\ H_2O$	0.23 mg
$CuCl_2 \cdot 2\ H_2O$	5.36 μg

Water to	100 ml
pH	7.4

[Reference: Personal communication from S. Goldstein, Biology Department, Brooklyn College, Brooklyn, New York.]

110. Isolation of Marine Fungi

Materials

Nutrient medium A	
Glucose	0.1 g
Yeast extract (or phytone)	0.01 g
Agar	1.8 g
Agar seawater* to	100 ml
pH (after autoclaving)	7.4–7.8
Nutrient medium B	
Peptone	0.01 g
Glucose	0.1 g
K_2HPO_4	0.005 g
Ferric citrate	0.001 g
Agar	1.8 g
Aged seawater* to	100 ml

Procedure

1. Wood scrapings, spore or mycelial masses, water samples, marine mud, or pieces of infected marine organisms are streaked over surface of either of the above media.
2. Incubate media in dark at 20°–30°C for one to several weeks.

Observation

Look for characteristic colonies or mycelia of yeasts and fungi. [Reference: T. W. Johnson, Jr., and F. K. Sparrow, Jr. "Fungi in Oceans and Estuaries," pp. 25–36. Cramer, Weinheim, 1961.]

111. Isolation of Marine Fungi

Materials

Nutrient medium	
Gelatin hydrolyzate	0.1 g
Glucose (add aseptically)	0.1 g

*Aged seawater is prepared by filtering freshwater to remove particles and storing it in the dark in glass carboys for 3–6 weeks.

Liver concentrate (1:20)	1.0 mg
Vitamin mixture (See Tables VII-9, VII-10, and VII-11.)	1.0 ml
Agar	1.5 g
Seawater	80 ml
Water to	100 ml

Penicillin G, 20,000 units/ml
Dihydrostreptomycin 5.0 mg/ml
Bent glass rod spreader (L-shape)

Procedure

1. Prepare agar nutrient medium above in Petri dishes.
2. Dry agar overnight but do not keep agar for more than two days.
3. Just before use spread surface of the agar with 2000 units of sterile penicillin and 0.5 mg sterile dihydrostreptomycin in concentrated aqueous solution.
4. Spread 0.2 ml of water sample over surface of nutrient agar medium and incubate in dark below 20°C for 7–10 days.

Observation

Look for colonies of fungi visible to the naked eye on low power magnification. [Reference: H. S. Vishniac, *Biol. Bull.* **111**, 410–14 (1956).]

112. Isolation of Aquatic Fungi Using a Millipore Filter

Materials

Low nutrient medium

Glucose	5.0 mg
Peptone	5.0 mg
Yeast extract	5.0 mg
Agar	3.0 g
Streptomycin sulfate (add aseptically)	0.05–0.1 g
Na penicillin G	0.05–0.1 g
Water (or seawater) to	100 ml

Millipore filter and membranes of small pore size

Procedure

1. Filter several liters of water sample through sterile membrane in filter.
2. Resuspend material on filter in 0.5 ml of sterile water and streak on low nutrient medium agar.
3. Incubate at 20°–25°C for several days.

Observation

Examine colonies on agar for presence of fungi. [Reference: C. E. Miller, *ASB Bull.* **13**, 42 (1966).]

113. Isolation of Aquatic Fungi

Materials

Hemp seeds (achenes of *Canabis sativa* L.)
Young grass leaves, cut in 1-inch lengths
Cellophane, cut in 1-inch squares
Nutrient media (add agar, 2 g/100 ml, of liquid medium to solidify)

(*a*) Yeast-starch

Yeast extract	0.4 g
Soluble starch	1.5 g
K_2HPO_4	0.1 g
$MgSO_4 \cdot 7\ H_2O$	0.05 g
Water to	100 ml

(*b*) Yeast-glucose

Yeast extract	0.1 g
Glucose	0.3 g
KH_2PO_4	0.14 g
Na_2HPO_4	0.06 g
$MgSO_4 \cdot 7\ H_2O$	0.01 g
Water to	100 ml

(*c*) Tryptone-glucose

Tryptone (or peptone)	1.0 g
Glucose	1.0 g
Water to	100 ml

(d) Cornmeal agar (BBL or Difco)

Procedure

1. Add two seeds (or pieces of leaf or a piece of leaf and cellophane) to 150 ml of filtered pond water in dish, a beaker covered by glass, or Petri dish.
2. Boil 10–20 minutes.
3. Cool quickly. This serves as nutrient ("bait") for aquatic fungi.
4. Add water or soil sample.
5. Incubate at room temperature in the dark for one to several weeks.

Observation

Examine culture for mycelium or hyphae or a drop of medium under the microscope (400–1000 ×) for motile zoospores. Zoospores are collected with a sterile capillary Pasteur pipette and a drop of water inoculated on any one of the above media solidified with agar. Spread drop on agar with a sterile L-shaped rod. Hyphae or mycelia may be transferred to any one of the above agar media with a sterile needle or loop.

Pure cultures may be obtained by repeated transfer onto sterile media. Antibiotics, i.e., penicillin (0.05%) or streptomycin (0.05%) may be added to medium aseptically to inhibit the growth of bacteria. [Reference: R. Emerson, *Mycologia* **50**, 589-621 (1958).]

114. Isolation of Cellulolytic Fungi

Materials

Czapek salt solution

$NaNO_3$	0.2 g
K_2HPO_4	0.1 g
$MgSO_4 \cdot 7\ H_2O$	0.05 g
KCl	0.05 g
$FeSO_4 \cdot 7\ H_2O$	1.0 mg
Water to	100 ml
pH	7.0-7.3

Filter paper in small strips
Concentrated HCl (sp. gr. 1.18)
Ethanol (95%)
Lactic acid
Büchner funnel

Procedure

1. Immerse 3.0 grams of filter paper strips in 100 ml concentrated HCl for 3 hours at 25°-27°C.
2. Shake occasionally.
3. Pour mixture with stirring into excess distilled water and allow to stand for 2-3 hours.
4. Pour off supernatant and wash residue on a Büchner funnel until water washes are no longer acid.
5. Wash once with 100 ml ethanol and dry in air. This is cellulose.
6. Soak cellulose in a few milliliters of Czapek salt solution, mix

well into a fine paste, and bring volume to 100 ml with Czapek salt solution.

7. Add agar (1.5 g) to harden medium.
8. Lactic acid may be added to bring pH to 5.4 and to inhibit bacterial growth.
9. Sterilize 20 ml in a tube at 15 psi. for 15 minutes.
10. Streak on surface plate or suspend soil or water sample in melted agar medium before pouring plate.
11. Incubate 4-6 days at 30°C.

Observation

Cellulolytic fungi may be recognized by the clear transparent zone around their colonies. [Reference: R. G. Bose, *Nature* **198**, 505-6 (1963).]

115. Enrichment and Isolation of Aquatic Fungi on Cellulose

Materials

Lens paper
Cellophane

Procedure

1. Inoculate soil or water into sterile pond or sterile seawater in a Petri dish containing small pieces of cellophane or lens paper floating on surface.
2. Incubate for several weeks at room temperature in the dark.

Observation

Examine paper or cellophane periodically under binocular dissecting microscope for the appearance of zoospores, mycelia, or thalli. Axenic cultures may be prepared by carefully isolating zoospores or a portion of mycelium and transferring this material to sterile water containing sterile cellophane or lens paper. Cultures growing on small squares of cellophane or lens paper lend themselves to study under high magnifications as a wet mount. [Reference: R. H. Haskins, *Am. J. Botany* **26**, 635-39 (1939).]

116. Enrichment and Isolation of Cellulolytic Fungi—Rautela and Cowling. See method No. 17.

117. Enrichment and Isolation of Aquatic Fungi, Chytrids

Materials

Uncoated cellophane
Cellulose agar medium (see procedures for cellulose agar plates, Method No. 17, 114)
Nutrient medium

Glucose (may be added aseptically)	0.5 g
Tryptone	0.4 g
Yeast extract	0.1 g
$MgSO_4 \cdot 7\,H_2O$	0.05 g
K_2HPO_4	0.1 g
H_3BO_3	0.3 mg
$CuSO_4 \cdot 5\,H_2O$	0.1 mg
H_3MoO_4(84%)	0.25 mg
$MnSO_4 \cdot 5\,H_2O$	2.5 mg
$ZnSO_4 \cdot 7\,H_2O$	4.0 mg
$FeCl_3 \cdot 6\,H_2O$	2.0 mg
$CaCl_2$	10.0 mg
Water to	100 ml
pH	6.4

Procedure and observation

Float uncoated cellophane squares (1 cm × 1 cm) on water sample. Incubate at 15°–25°C until a rhizomycelium forms on cellophane. Remove cellophane, wash with sterile water, and blot dry with sterile filter paper. Place cellophane on dry surface of cellulose agar medium and incubate for 10 days at 15°–25°C or until sporangia appear on agar surface. Cut out portion of agar with rhizomycelium and sporangia at some distance from cellophane inoculum with a sterile loop or needle and transfer to nutrient medium. Rhizomycelium develops in liquid nutrient medium. If rhizomycelium placed in water for 3–5 days, sporangia and zoospores are produced. [Reference: S. Goldstein, *Am. J. Botany* 48, 294–8 (1961).]

118. Enrichment and Isolation of the Rotifer-Trapping Fungus, *Zoophagus*

Materials

Activated sludge from sewage system or pond water or mud containing rotifers
Cornmeal agar

Procedure

1. Inoculate each of several cornmeal agar plates with 1 ml of sludge or pond water or mud and 5 ml of distilled water. This produces a semifluid medium on the agar surface.
2. Incubate for 1 week at 25°C.

Observation

After one week examine semifluid material in a wet mount under microscope (400 ×). Look for rotifers (occasionally nematode worms) trapped in hyphal pegs in a fungal mycelium made up of short filaments. The fungus produces gemmae in groups of 2 to 12. The fungus may be isolated free of most contaminating microorganisms by transferring the rotifers and gemmae to fresh cornmeal agar moistened with 5 ml of sterile water. [Reference: W. O. Pipes, and D. Jenkins, *Intern. J. Air Water Pollution* 9, 495–500 (1965).]

119. ISOLATION OF MARINE YEASTS

Materials

Isolation medium*	
Glucose	2.0 g
Peptone	1.0 g
Yeast extract	0.5 g
Agar	2.0 g
Filtered seawater to	100 ml
pH (adjusted with lactic acid)	4.5

Procedure

1. Filter seawater and place filter disc on surface of medium.
2. Incubate 2–3 days at 18°–25°C in the dark.

Observation

Look for colonies and examine colony microscopically for yeasts. [Reference: J. W. Fell, and N. van Uden, *in* "Symposium on Marine Microbiology" (C. H. Oppenheimer, ed.), pp. 329-340. Thomas, Springfield, Illinois, 1963.]

*Bacteria may be inhibited by antibiotic mixture: chlortetracycline · HCl, 1.0 mg%, chloramphenicol, 2.0 mg%, streptomycin · SO_4, 2.0 mg%, added aseptically.

120. Isolation of Marine Nitrogen-Fixing Yeasts

Materials

Millipore filter

Nutrient medium* (can be hardened with agar, 1.5 g)

NaCl	2.92 g
$MgSO_4 \cdot 7\ H_2O$	1.23 g
$CaCl_2$	0.11 g
KCl	0.075 g
K_2HPO_4	0.017 g
Fe	0.4 mg
Mn	0.05 mg
B	0.05 mg
Zn	5.0 μg
Cu	2.0 μg
Mo	1.0 μg
V	1.0 μg
Co	1.0 μg
Water to	100 ml
pH	7.6–7.8

Procedure

1. Sterilize 100 ml of nutrient medium in 500-ml flask.
2. Filter large seawater sample through Millipore filter (0.45 μ pore size).
3. Place filter disc in liquid medium or on agar plate containing nutrient medium.
4. Incubate for 1–4 months in dark at 15°–25°C for pink nitrogen-fixing *Rhodotorula* yeast.

Observation

Look for growth of yeast or nitrogen-fixing bacteria in medium lacking inorganic or organic nitrogen. [Reference: M. B. Allen, *In* "Symposium on Marine Microbiology" (C. H. Oppenheimer, ed.), pp. 85–92. Thomas, Springfield, Illinois, 1963.]

*Medium may be supplemented with: thiamine, 10 μg/100 ml; biotin, 1.0 μg/100 ml; Vitamin B_{12}, 0.5 μg/100 ml.

121. Isolation of Yeast from Freshwater

Materials

Isolation medium

Glucose	2.0 g
Peptone	1.0 g
Yeast extract	0.5 g
Agar	2.0 g
Water to	100 ml
pH (adjusted with lactic acid)	4.5

Millipore filter

Millipore filter membrane (HA) 0.45 μ porosity

Procedure

1. Filter water sample through membrane.
2. Transfer membrane to isolation medium in Petri dish.
3. Incubate at 18°-20°C. for 3-5 days.
4. Subculture yeasts from membrane to periphery of plate.

Observation

Examine cultures microscopically for presence of yeasts.

[Reference: N. van Uden, and D. C. Ahearn, *Antonie van Leeuwenhoek J. Microbiol. Serol.* **29**, 308-312 (1963).]

122. Isolation of Protostelids, Mycetozoa (Myxomycetes)

Materials

Source of organisms: soil, humus, dung, rotting wood, and dead plant parts

Isolation media

(*a*) *Lactose*	0.1 g
Yeast extract	0.05 g
Agar	2.0 g
Water to	100 ml
(*b*) *Weak hay infusion*	
Hay	0.25 g
K_2HPO_4	0.2 g
Agar	2.0 g

Water to	100 ml
Escherichia coli cells	
Aerobacter aerogenes cells	

Procedure

1. Place soil or other source material in small portions on about 8 well-scattered sites on isolation agar medium in Petri dish. Soil may also be sprinkled on medium.
2. Incubate at room temperature.
3. After 3 days of incubation observe daily with a dissecting binocular microscope for 8–9 days.

Observation

Look for uninucleate ameboid protoplasts; some are multinucleate and may form a reticulate plasmodium. Some species have 1–2 flagella. Sporocarp (spore stalk) noncellular, short, and slender with 1 and sometimes 2 spores at apex. Cultures may be maintained by transferring spores or ameba to fresh isolation medium along with *E. coli* or *A. aerogenes* cells or cells isolated with the protostelids. [Reference: L. S. Olive, *Mycologia* **59**, 1–29 (1967).]

123. Isolation of Slime Molds

Materials

Sterile mortar and pestle
Sterile water
Dilute nutrient agar media

(*a*) Dilute hay-infusion agar (¼ strength hay-infusion agar)

Heat 3.5 g of partially decomposed hay in 100 ml of water for one-half hour at 110°C and 10–15 psi. Filter. Add K_2HPO_4, 0.2 g, and adjust volume of filtrate to 400 ml. Add agar 1.5% and adjust pH to 6.0–6.2.

(*b*) Peptone-sugar agar

Peptone	0.05–0.1 g
Lactose (or glucose)	0.05–0.1 g
Agar	1.5–2.0 g
Water to	100 ml
pH	6–7

(*c*) Carrot-infusion agar

Boil 30 g of fresh carrots in 100 ml of water. Filter. Add agar (1.5–2.0 g) to filtrate and bring volume to 100 ml. Adjust pH to 6–7.

Escherichia coli culture

Rich media for maintaining slime molds

(*a*) Peptone-yeast extract agar

Peptone	0.2 g
Yeast extract	0.2 g
Glucose	0.5 g
Agar	1.5–2.0 g
Water to	100 ml
pH	6–7

(*b*) Peptone-lactose agar

Peptone	1.0 g
Lactose	1.0 g
Agar	1.5–2.0 g
Water to	100 ml
pH	6–7

(*c*) Enriched hay-infusion agar

Peptone	0.5 g
Hay-infusion agar*	100 ml

(*d*) Canned pea broth agar

Canned pea broth	10.0 ml
Peptone	0.5 g
Phosphate buffer	0.02 *M*
Agar	1.5–2.0 g
Water to	100 ml
pH	7.0

(*e*) Bonner's agar

Peptone	1.0 g
Glucose	1.0 g
Na_2HPO_4	0.096 g
KH_2PO_4	0.145 g
Agar	2.0 g
Water to	100 ml
pH	6.0

Procedure

1. Sample (soil, leaf mould, etc.) is ground in sterile mortar with an equal volume of sterile water.

*3.5 g of partially decomposed hay heated in 100 ml of water for one-half hour at 110°C and 10–15 psi. Filter. Add K_2HPO_4, 0.2 g, and adjust volume to 100 ml.

2. This suspension is then streaked on any one of the depleted agar media listed. The use of media with little organic material minimizes the overgrowth of the slime molds by fungi or bacteria.
3. Incubate plates for 2–3 weeks at 16°–18°C.

Observation

Examine plates for wheel-like pseudoplasmodium (Fig. VII-13). Pseudoplasmodium may aggregate and develop fruiting structures (Fig. VII-14). Use low power (100 ×) to examine plate surface. Pure cultures may be obtained by transferring fruiting structure (spores) from elevated structure and transplanting to one of the dilute agar media listed, upon which *Escherichia coli* have been previously grown for several days. An alternative procedure is to remove a portion of pseudoplasmodium with a sterile loop or spatula, avoiding contaminating fungi as much as possible, and transferring pseudoplasmodium to a streak of *E. coli* previously grown on any one of the dilute agar media. Purified slime mold cultures may be maintained on any one of the rich media listed above upon which *E. coli* are first grown or heat-killed *E. coli* cells streaked on surface. [Reference: K. B. Raper, *Quart. Rev. Biol.* **26**, 169–90 (1961).]

124. Counting Mold Populations by Microplating*

Materials

Nutrient agar medium for molds
Clean slides
Stain: phloxine (1%) and fast green (1%) dissolved in an aqueous solution of phenol (5%)

Procedure

1. Dilute sample (soil suspended in sterile water or water).
2. Add 1 drop of sample to 1–2 drops of melted agar medium mix and spread over limited area of slide (1–2 cm) with an L-shaped inoculating needle.
3. Harden and incubate in a moist chamber (any closed container with moist blotting or filter paper supplying moisture) for 24–30 hours at 28°C.
4. Dry slide over a hot plate at 85°C for 5–8 minutes. Slides may be stored until stained.
5. Stain dry slide by immersing it in stain for 2–3 seconds.
6. Wash off excess dye.

*This method may be adapted for other microorganisms

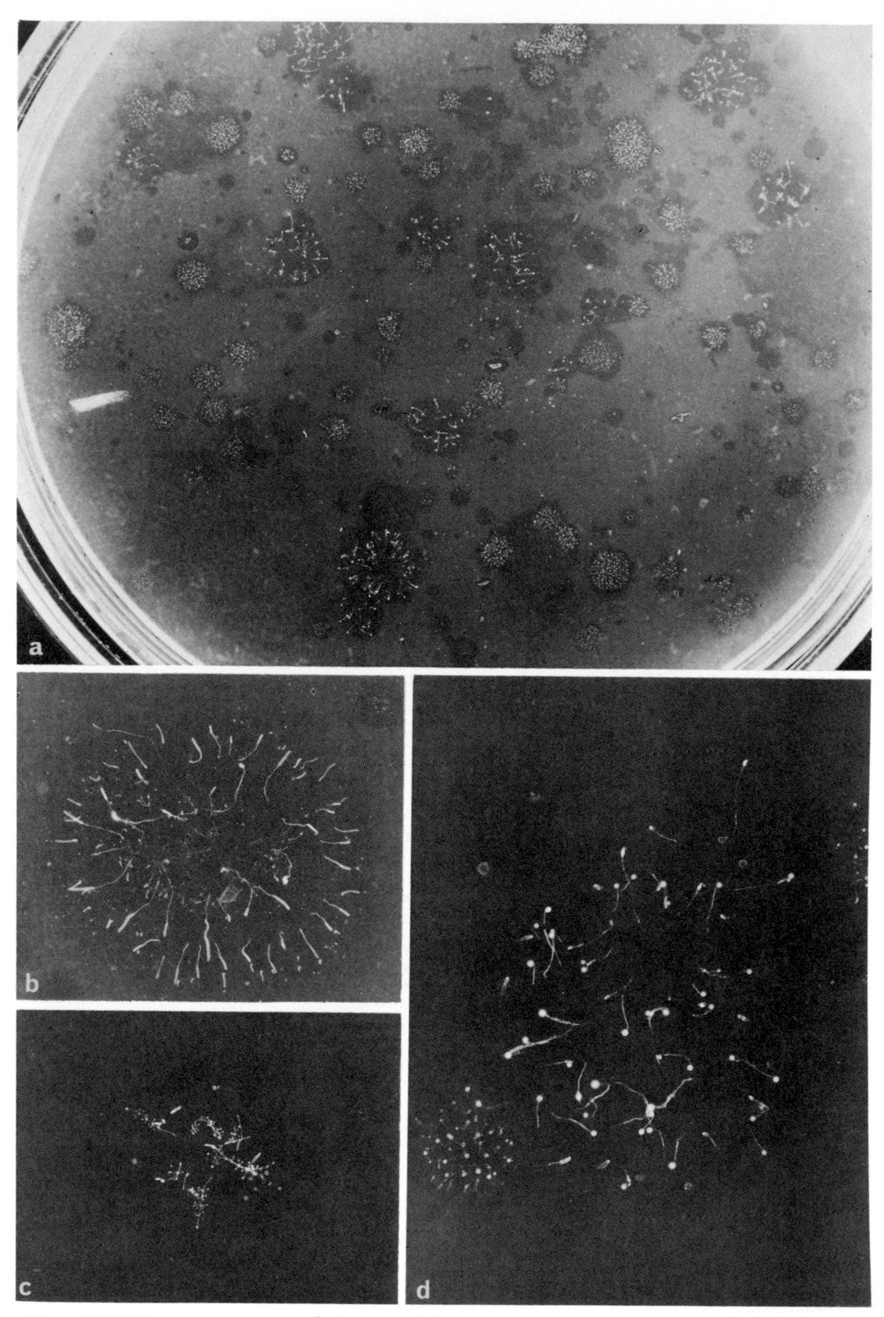

FIG. VII-13. Appearance of slime molds colonies on agar. [From Cavender, J., and Raper, K. B., *Am. J. Botany* **52**, 294-296 (1965).] (a). Isolation plate showing cellular slime mold clones. Clear areas (plaques) without fruiting structures result from myxamebae or free-living amebae eating the bacterial lawn. (b). *Polysiphonia violaceum* colony, (c). *Polysiphonia pallidum* colony. (d). A large colony of *Dictyostelium mucoroides* and a colony of the much smaller species *D. minutum* in the upper right corner of the figure.

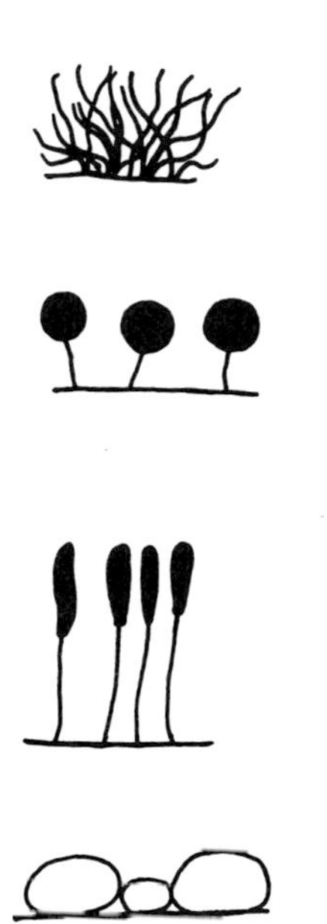

FIG. VII-14. Slime mold fruiting structures. Top to bottom: *Ceratiomyxa*, *Physarum*, *Stemonitis*, and *Lycogala*.

Observation

Colonies of molds are blue or dark purple against pale background when examined under a binocular dissecting microscope (100 ×). [Reference: J. E. Roberts, and W. B. Bollen, *Appl. Microbiol.* **3**, 190–4 (1955).]

IV. Algae

125. Isolation and Culture of Blue-green Algae

Materials

Algal material from soil, pond, stream or lake

Nutrient medium

$Ca(NO_3)_2 \cdot 4\ H_2O$	0.004 g
K_2HPO_4	0.001 g
$MgSO_4 \cdot 7\ H_2O$	0.0025 g
Na_2CO_3	0.002 g
$Na_2SiO_3 \cdot 9\ H_2O$	0.0025 g
Ferric citrate	0.3 mg
Citric acid	0.3 mg

Water to	100 ml
pH	8.0-9.5

Procedure and Observation

Inoculate algal material in 50 ml of above nutrient medium in a flask. Incubate in white light at 25°C for several weeks or until there is good growth. Reinoculate sterile medium with desired blue-green alga to minimize contaminants. Axenic cultures may be developed by exposing blue-green algae to varying intensities of ultraviolet light and then transferring algae to sterile medium. Some cultures will develop free of bacterial and other contaminants which are more sensitive to ultraviolet. Many algae will also die. (See Fig. VII-15.) [Reference: G. C. Gerloff, G. P. Fitzgerald, and F. Skoog, *Am. J. Botany* **37**, 216-8 (1950).]

126. Method for Obtaining Bacteria-Free Cultures of Blue-Green Algae

Materials

Neutral to slightly alkaline soils and water samples

Enrichment medium lacking combined nitrogen

K_2HPO_4	0.5 g
$MgSO_4 \cdot 7\ H_2O$	0.5 g
Na_2CO_3	0.05 g
$CaCl_2$	0.05 g
Fe (as Fe-EDTA)	5.0 mg
Trace elements*	1.0 ml
Water to	100 ml
pH	7.0-7.4

Yeast extract-glucose agar plates

Silica gel plates or agar plates of the enrichment medium

Procedure

1. Inoculate soil or water sample into 500-ml flask containing 200 ml of the enrichment medium and incubate in fluorescent light for 2-3 weeks.
2. When algae are numerous take 5.0 ml of a thick suspension in a tube and place tube in a water bath at 47°-48°C.

*Trace element solution: H_3BO_3, 0.31 g; $MnSO_4 \cdot H_2O$, 0.223 g; $ZnSO_4 \cdot 7\ H_2O$, 28.7 mg; $Na_2MoO_4 \cdot 2\ H_2O$, 8.8 mg; $CuSO_4 \cdot 5\ H_2O$, 12.5 mg; $Co(NO_3)_2 \cdot 6\ H_2O$, 14.6 mg; KBr, 11.9 mg; KI, 8.3 mg; water to 100 ml.

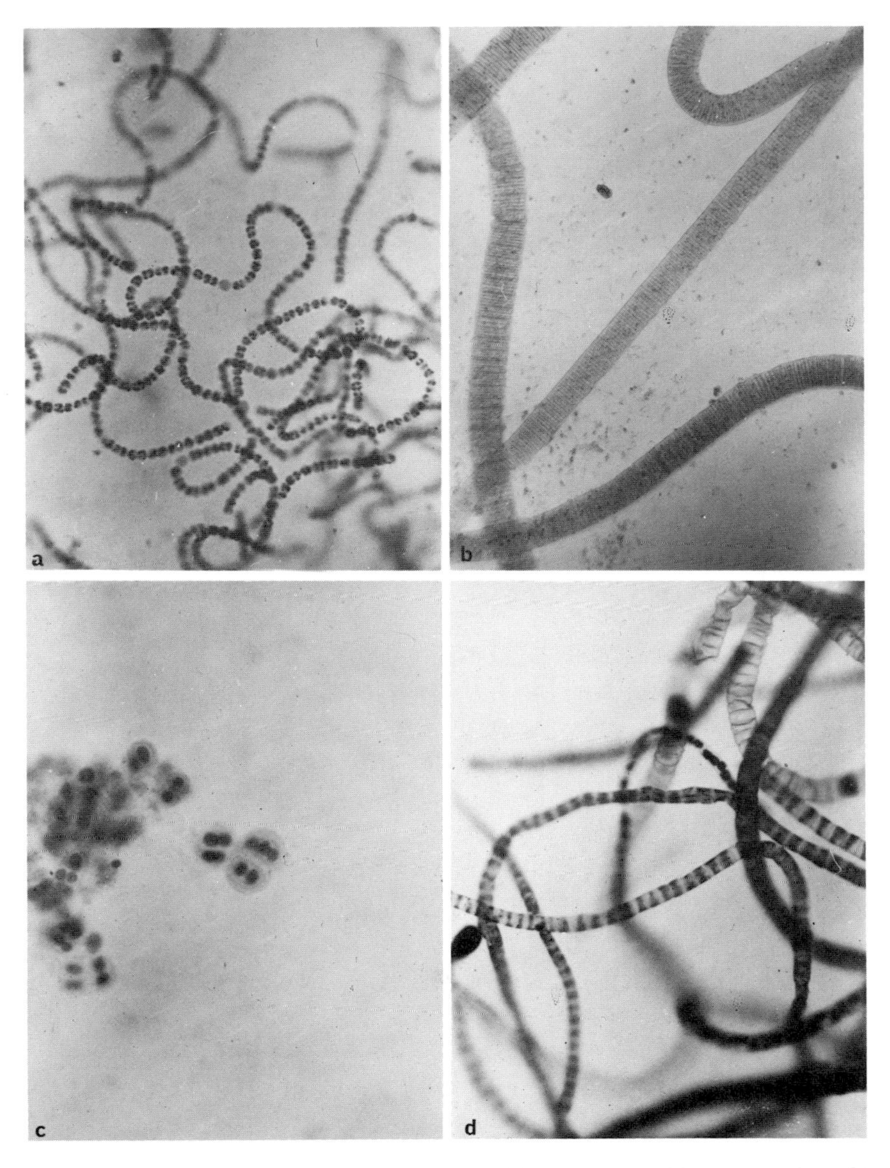

FIG. VII-15. Some common algae, stained. (a). *Nostoc* sp., blue-green alga; (b). *Oscillatoria* sp., blue-green alga; (c). *Gloeocapsa* sp., blue-green alga; (d). *Ulothrix* sp., green alga; (e). *Spirogyra* sp., green alga; (f). desmid; (g). desmid; (h). desmid; (i). *Anabaena* sp., blue-green alga; (j). *Aphanizomenon flos-aquae* (L.) *Ralfs*, blue-green alga; (k). *Calothrix* sp., blue-green alga; (l). *Microcoleus* sp., blue-green alga; (m). *Lyngbya* sp., blue-green alga; (n). *Synura uvella Ehr.*, chrysophyte alga. (i-n from Prescott, G. W. "How to Know the Fresh-water Algae" W. C. Brown, Dubuque, Iowa, 1964).

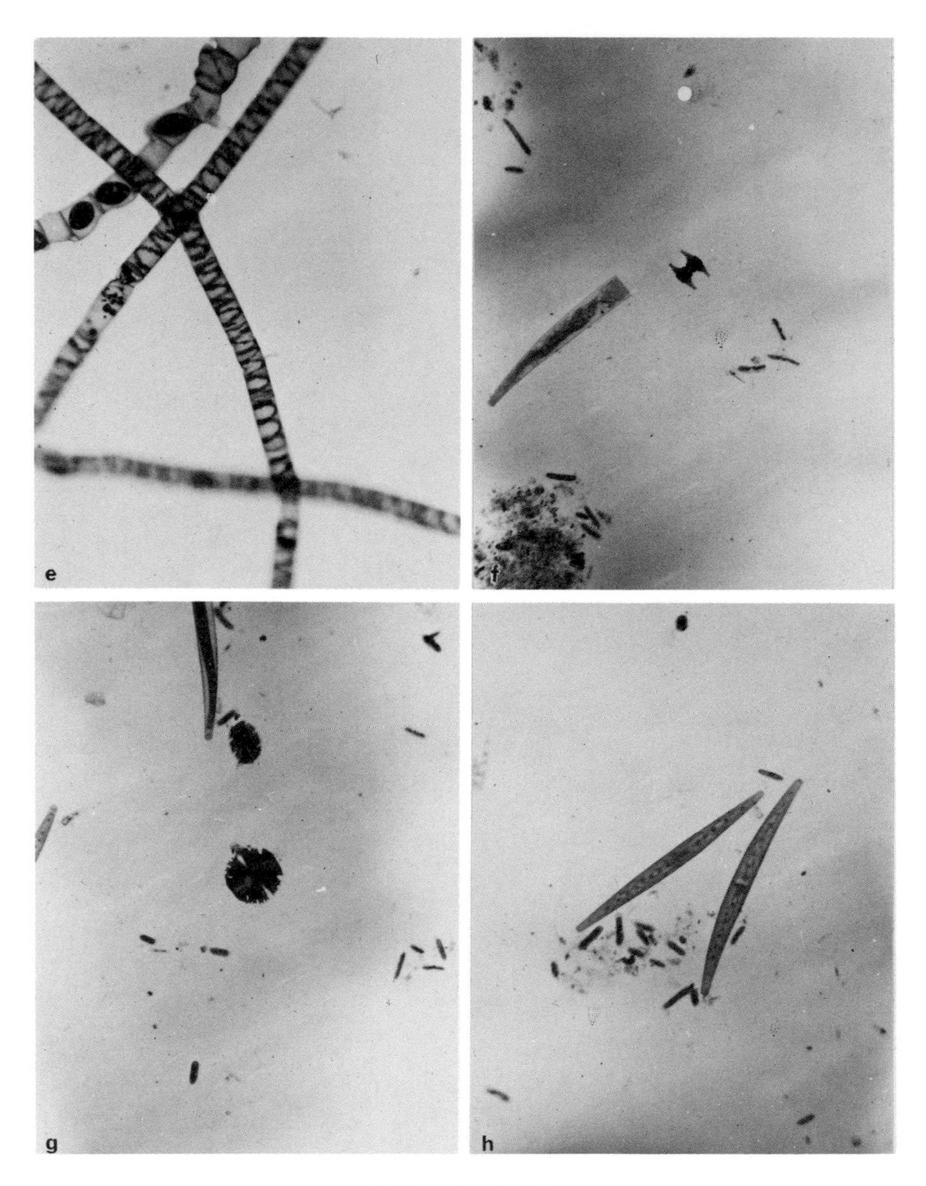

FIG. VII-15e,f,g,h

3. After each of 15, 30, 45, 60, and 75 minutes of heating a drop of culture is inoculated into each of the following: 10 ml of enrichment medium in screw-top flasks or tubes, streaked onto agar or silica gel plates, and yeast extract-glucose plates.

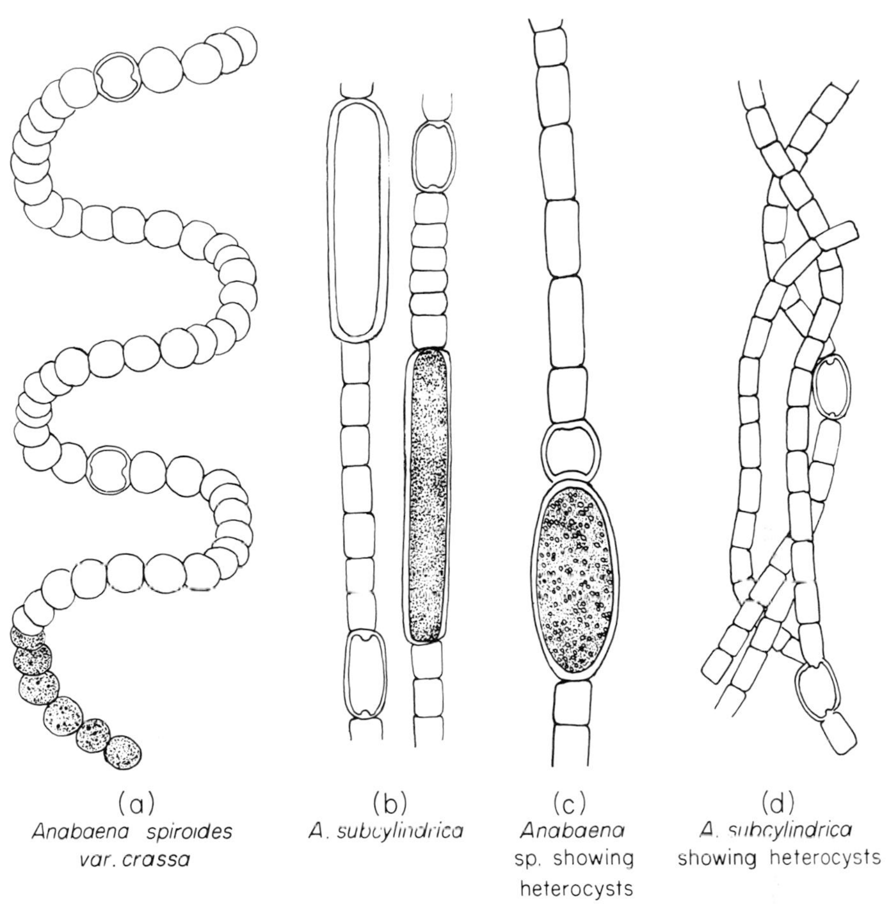

FIG. VII-15i

Observation

If no bacteria grow on yeast extract-glucose agar they are also likely to be absent from blue-green algal cultures. (See Fig. VII-15.) [Reference: K. T. Wieringa, *Antonie van Leeuwenhoek J. Microbiol. Serol.* **34**, 54–56 (1968).]

127. Isolation of Marine Nitrogen-Fixing Blue-Green Algae

Materials

Millipore filter
Nutrient medium*

*Medium may be supplemented with thiamine, 10 μg/100 ml, biotin, 1.0 μg/100 ml, Vitamin B_{12}, 0.5 μg/100 ml.

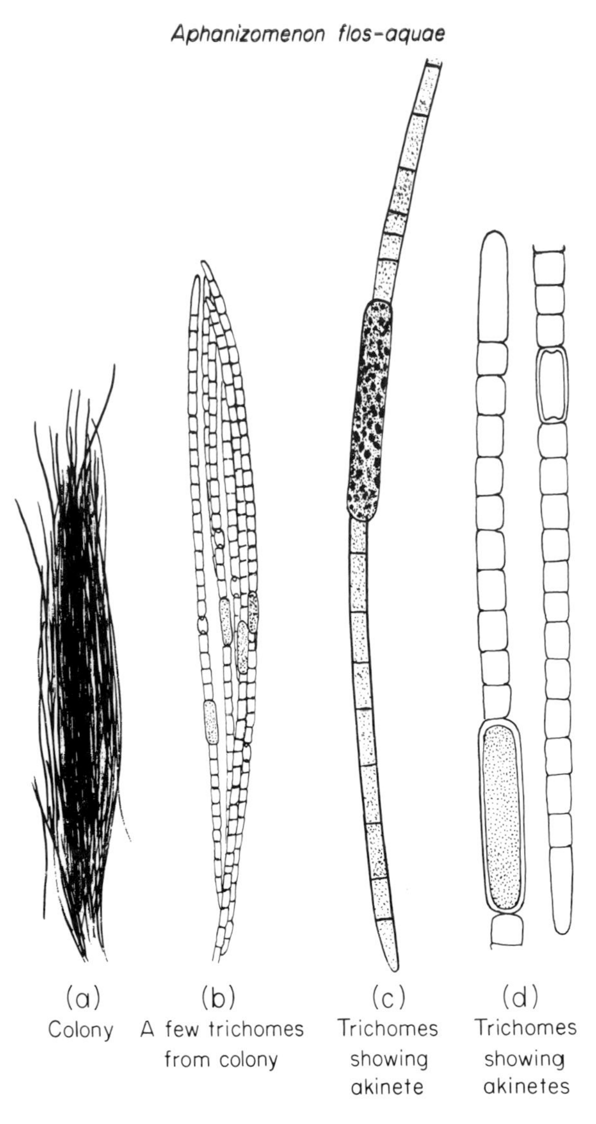

Fig. VII-15j

(Medium may be hardened with Ionagar, 1.5%)

NaCl 2.92 g
$MgSO_4 \cdot 7\,H_2O$ 1.23 g
$CaCl_2$ 0.11 g
KCl 0.0745 g

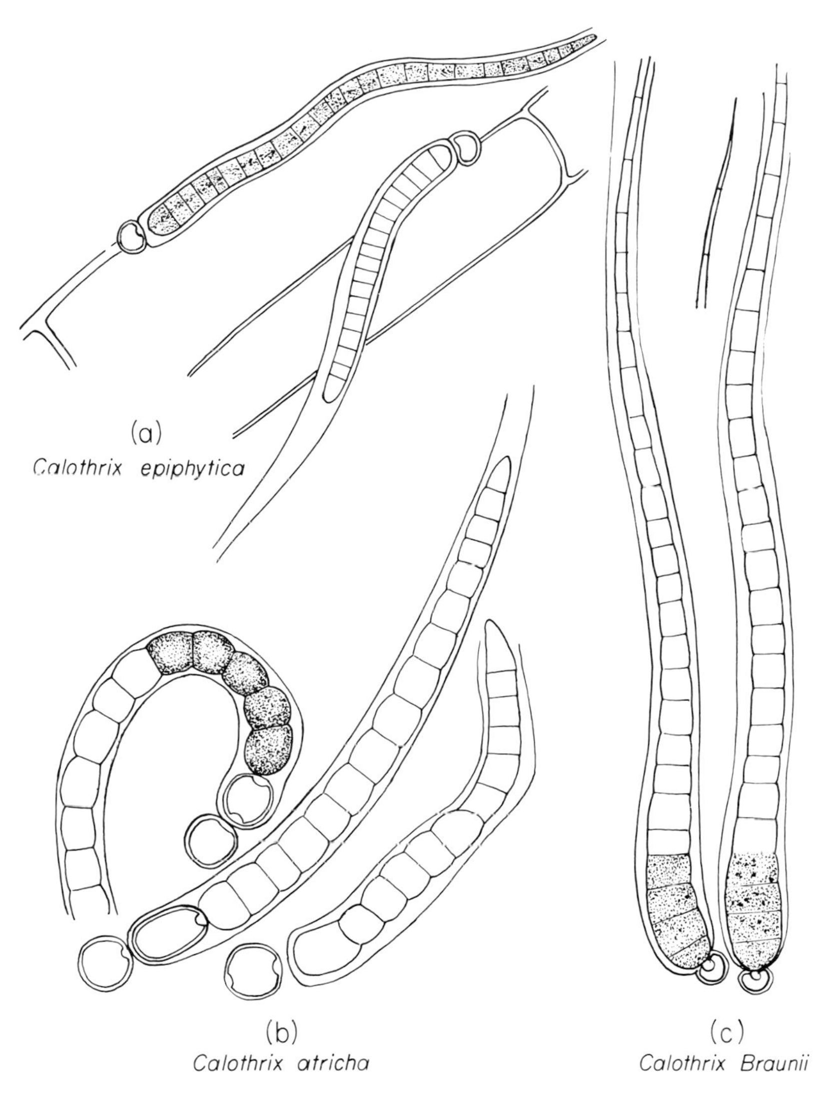

FIG. VII-15k

K_2HPO_4	0.0174 g
Fe	0.4 mg
Mn	0.05 mg
B	0.05 mg
Zn	5.0 μg
Cu	2.0 μg

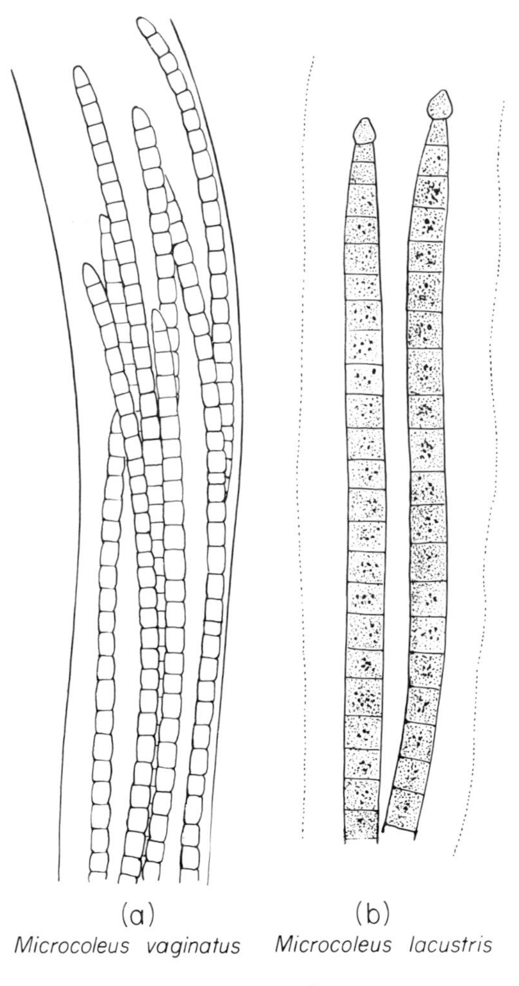

FIG. VII-15l

Mo	1.0 μg
V	1.0 μg
Co	1.0 μg
Water to	100 ml
pH	7.6–7.8

Procedure

1. Sterilize 100 ml above medium in 500-ml flask.
2. Filter large seawater sample through Millipore filter (0.45 μ pore size).

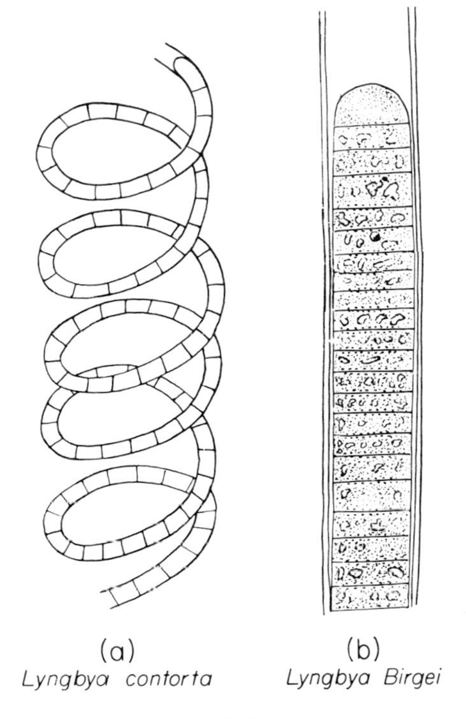

FIG. VII-15m

3. Place filter disc in liquid medium or on agar plate containing nutrient medium.
4. Incubate at 15°-25°C for 1-4 months in light for nitrogen-fixing blue-green algae.

Observation

Look for growth of algae in medium lacking inorganic or organic nitrogen. (See Fig. VII-15.) [Reference: M. B. Allen, 1963. *In* "Symposium on Marine Microbiology" (C. H. Oppenheimer, ed.) pp. 85-92. Thomas, Springfield, Illinois, 1963.]

Note: All of these media may show precipitates, particularly iron salt precipitates. These may be ignored or a chelating agent [citric acid or its salts (0.05%) or ethylenediaminetetraacetic acid (0.05%)] may be added to solubilize the precipitate. Media may also be supplemented with organic material [soil extract (0.001-0.005%), trypticase (0.001-0.005%), liver solution (0.005%) as well as trace metals.]

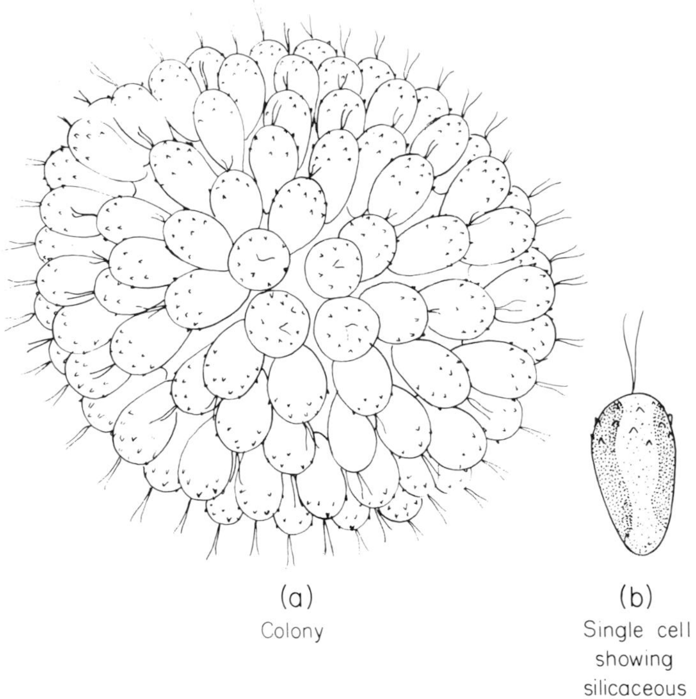

FIG. VII-15n

128. MEDIA FOR THE GROWTH AND ISOLATION OF ALGAE (PHYTOFLAGELLATES)*

Materials

Molisch medium (may be diluted 1:2 or 1:4)

$(NH_4)_2HPO_4$	0.08 g
K_2HPO_4	0.04 g
$MgSO_4 \cdot 7\ H_2O$	0.04 g
$CaSO_4$	0.04 g
K_2HPO_4	0.02 g
Water to	100 ml

*May be diluted. $Ca(NO_3)_2$ may be replaced by other Ca salts.

Beijerinck medium (best for algae from acid habitats)

NH_4NO_3	0.1 g
K_2HPO_4	0.02 g
$MgSO_4 \cdot 7\ H_2O$	0.01 g
$FeCl_3 \cdot 6\ H_2O$	0.1 mg
Water to	100 ml

Knop medium*

KNO_3	0.1 g
$Ca(NO_3)_2$	0.01 g
K_2HPO_4	0.02 g
$MgSO_4 \cdot 7\ H_2O$	0.01 g
$FeCl_3 \cdot 6\ H_2O$	0.1 mg
Water to	100 ml

Pringsheim medium

KNO_3	0.02 g
$(NH_4)_2HPO_4$	2.0 mg
$MgSO_4 \cdot 7\ H_2O$	1.0 mg
$CaCl_2 \cdot 6\ H_2O$	0.05 mg
$FeCl_3 \cdot 6\ H_2O$	0.05 mg
Water to	100 ml

Detmer medium (modified)

$Ca(NO_3)_2 \cdot 4H_2O$	0.03 g
KH_2PO_4	8.0 mg
KCl	8.0 mg
$MgSO_4 \cdot 7\ H_2O$	8.0 mg
$FeCl_3 \cdot 6\ H_2O$	10.0 mg
Water to	100 ml

Czurda medium

KNO_3	0.02 g
K_2HPO_4	0.002 g
$MgSO_4 \cdot 7\ H_2O$	0.001 g
$FeSO_4 \cdot 7\ H_2O$	0.5 mg
$CaSO_4$	0.2 mg
Water to	100 ml

Chu medium

$Ca(NO_3)_2 \cdot 4\ H_2O$	4.0 mg
K_2HPO_4	1.0 mg
$MgSO_4 \cdot 7\ H_2O$	25.0 mg
Na_2CO_3	2.0 mg

*May be diluted. $Ca(NO_3)_2$ may be replaced by other Ca salts.

$Na_2SiO_3 \cdot 9H_2O$	2.5 mg
$FeCl_3 \cdot 6\ H_2O$	0.08 mg
Water to	100 ml

Procedure

1. Inoculate sterile media in tubes or Petri dishes containing media hardened with agar 1.5-2.0%.
2. Incubate in subdued light (incandescent or fluorescent) for 1-3 weeks at 20°-30°C.

Observation

Growth of algae (or flagellates) in tubes or Petri dishes may be seen by eye or with the aid of low power objective of microscope. Transfer algal material to fresh sterile medium until free of contaminants. (See Figs. VII-15, VII-16, and VII-17.) [Reference: E. G. Pringsheim *In* "Manual of Phycology" (G. M. Smith, ed.), pp. 347-57. Chronica Botanica Co., Waltham, Massachusetts, 1951.]

129. MEDIA FOR THE GROWTH AND ISOLATION OF MARINE AND FRESHWATER ALGAE*

Materials

Cyanophycean (blue-green algae) agar	
KNO_3	0.5 g
K_2HPO_4	0.01 g
$MgSO_4 \cdot 7\ H_2O$	0.005 g
Fe NH_4 citrate	1 drop of 1% solution
Agar	1.5 g
Water to	100 ml
Desmid agar†	
$MgSO_4 \cdot 7\ H_2O$	1 mg
K_2HPO_4	1 mg
KNO_3	10 mg
Agar	0.75 g
Water to	100 ml
Soil-water medium*	
Fresh garden soil	¼-½ inch
Water to	¾ volume of container

*$CaCo_3$ may be added to make all media alkaline.

†Some desmids grow better if 5 ml/100 ml of soil-water medium supernatant is added.

*Steam mixture for one hour on two successive days. Use supernatant fluid.

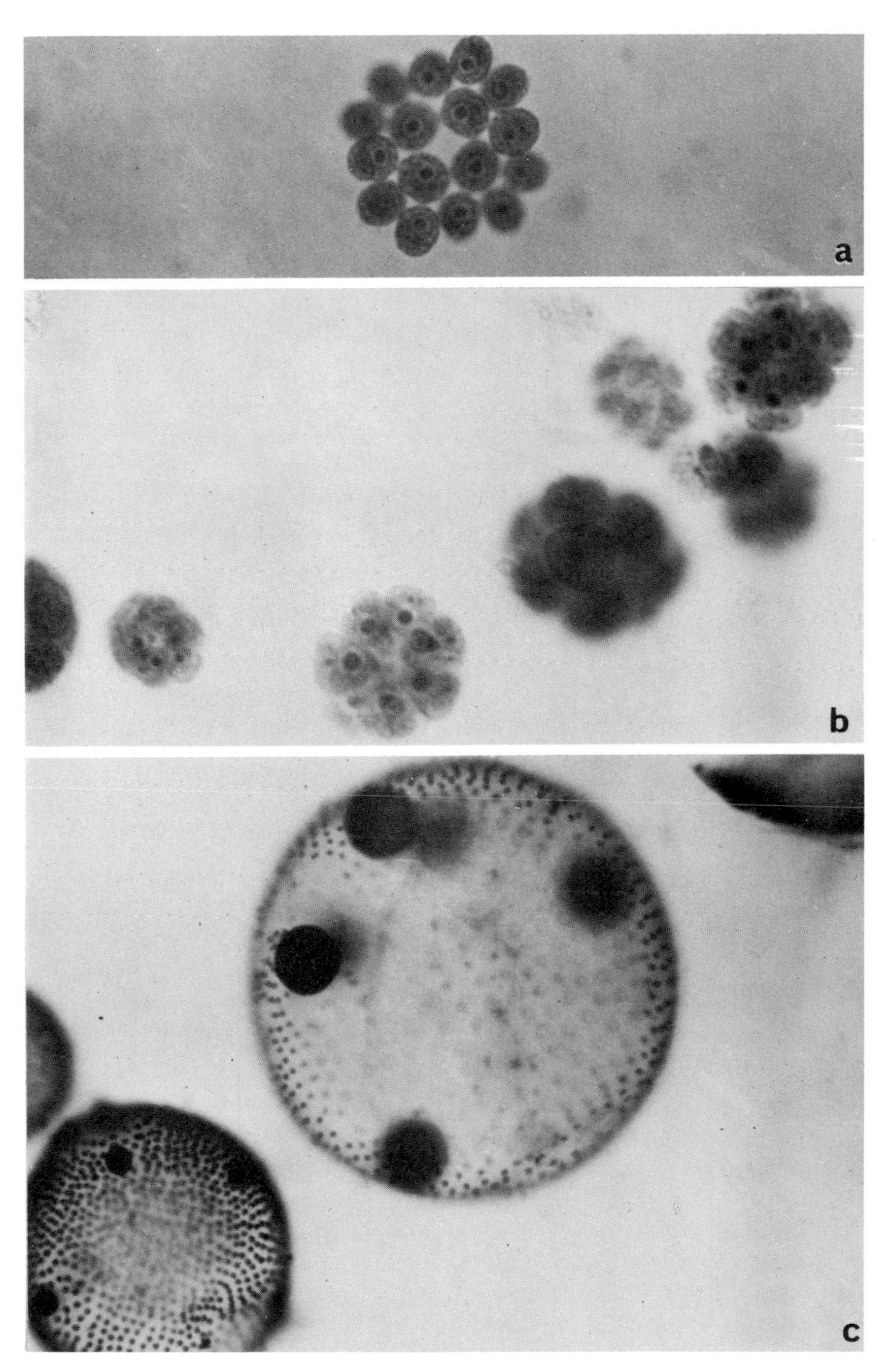

FIG. VII-16. Some common colonial phytoflagellates, stained. (a). *Gonium* sp.; (b). *Pandorina* sp.; (c). *Volvox* sp.

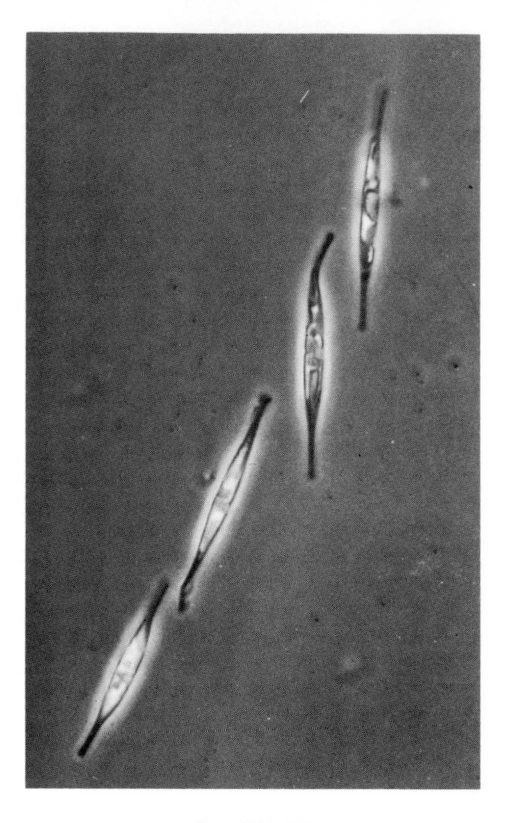

FIG. VII-17a

FIG. VII-17. Some common salt marsh and marine diatoms. (a). *Nitzschia acicularis;* (b). *Asterionella* sp.; (c). *Navicula digitoradiata*; (d). *Amphora proteus*; (e). *Grammatophora murina*; (f). *Bacillaria paradoxa*; (g). epiphytic diatom on alga; (h). *Biddulphia* sp.; (i). Cluster of *Asterionella*. (a–g, from Dr. J. J. Lee; h and i, from Dr. K. Gold.)

Bristol's medium. Separate solutions of each of the following:*

$NaNO_3$	2.5%
$CaCl_2$	0.25%
$MgSO_4 \cdot 7\,H_2O$	0.75%
K_2HPO_4	0.75%
KH_2PO_4	1.75%
NaCl	0.25%

*Add 10 ml of each of the above solutions to 940 ml water. Add 1 drop of 1% $FeCl_3$ solution. Sterilize. To prepare agar medium use agar 1.5%.

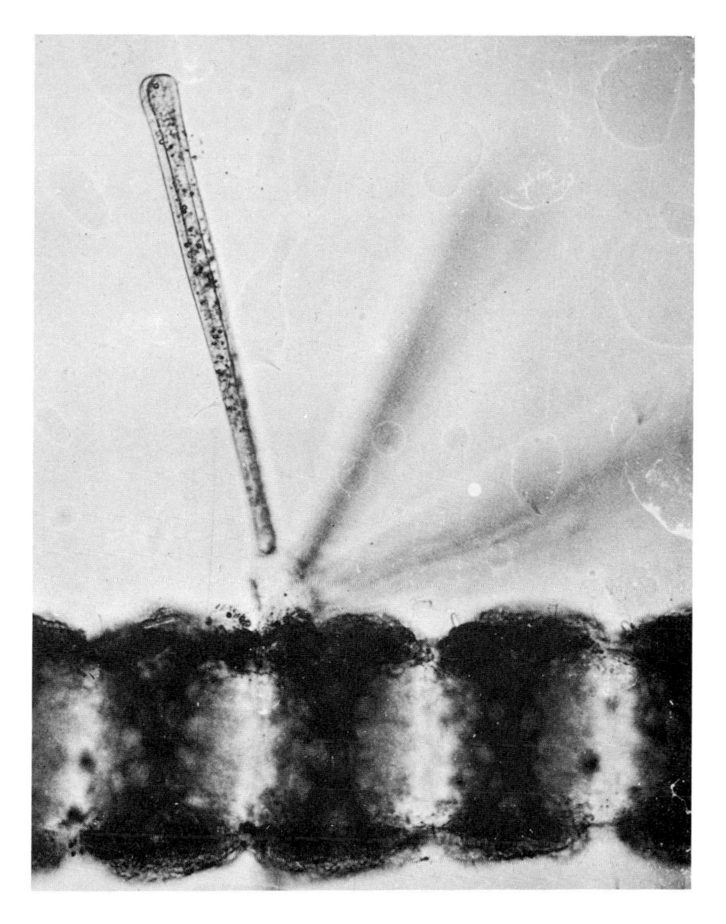

FIG. VII-17b

Erdschreiber solution‡	
Filtered seawater	100 ml
Soil-water medium supernatant	5.0 ml
$NaNO_3$	0.02 g
Na_2HPO_4	0.003 g
Malt agar	
Malt extract	7.5 g
Agar	1.5 g

‡*First day:* Filter seawater through No. 1 filter paper and then heat to 73°C; *Second Day:* Heat to 73°C again. Autoclave $NaNO_3$ and $NaHPO_4$ as 1.0 ml solutions; *Third day:* Add cold salt solutions to cold soil-water supernatant and then add to seawater and dispense in sterile tubes.

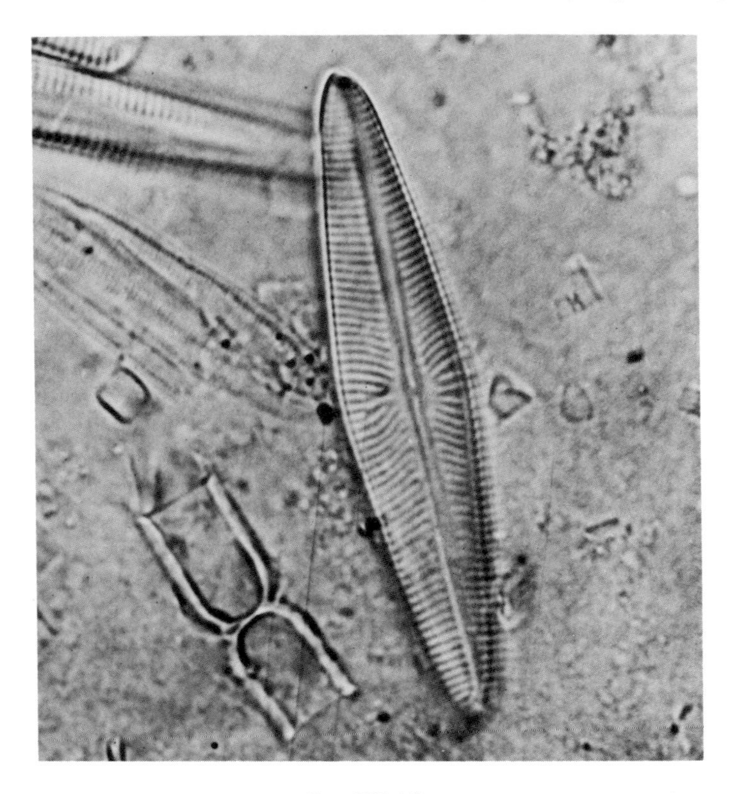

Fig. VII-17c

Water to	100 ml
NBB agar	
Sodium acetate · 3 H_2O	0.05 g
Beef extract	0.05 g
Tryptone	0.05 g
Agar	1.5 g
Water to	100 ml
Soil-extract agar	
Soil-water medium supernatant	4.0 ml
Bristol's medium	96.0 ml
Agar	1.5 g
Diatom seawater agar	
Bristol's solution	50 ml
Seawater	50 ml
Supernatant from soil-water medium	5 ml
Agar	1.5 g

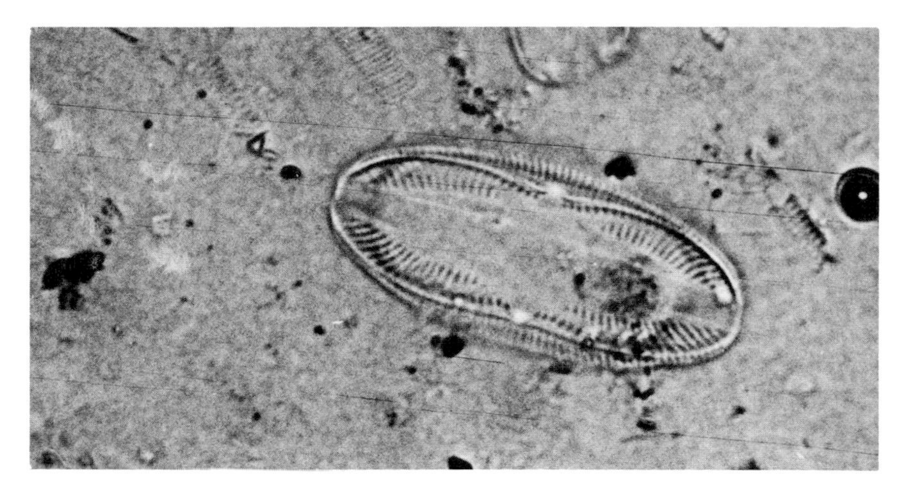

FIG. VII-17d

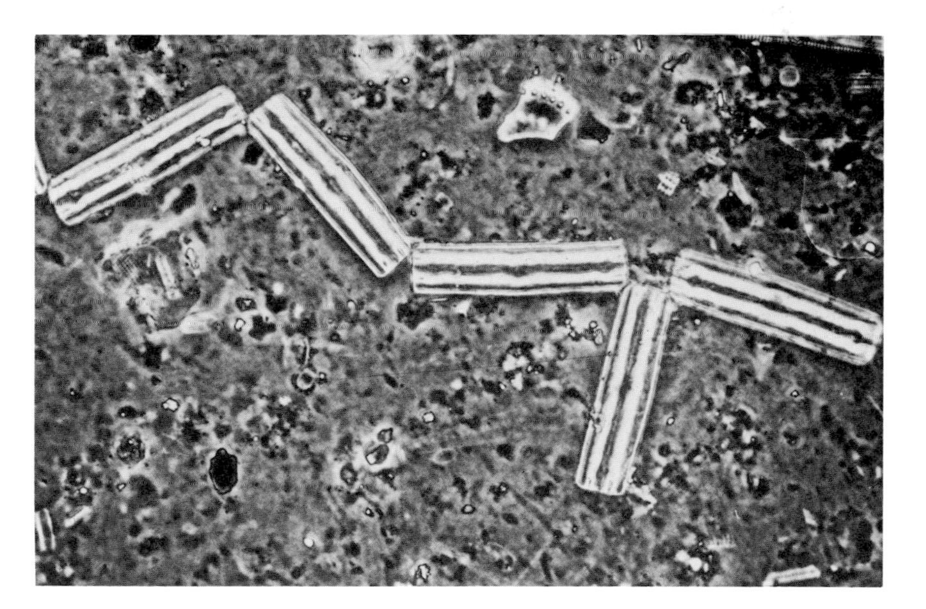

FIG. VII-17e

Procedure

1. Inoculate any one or more of the above sterile media (10 ml/tube or as solid medium in Petri dish) with soil or water sample.
2. Incubate in the light at 20°-30°C for 1-3 weeks. Cold water forms may require lower incubation temperatures.

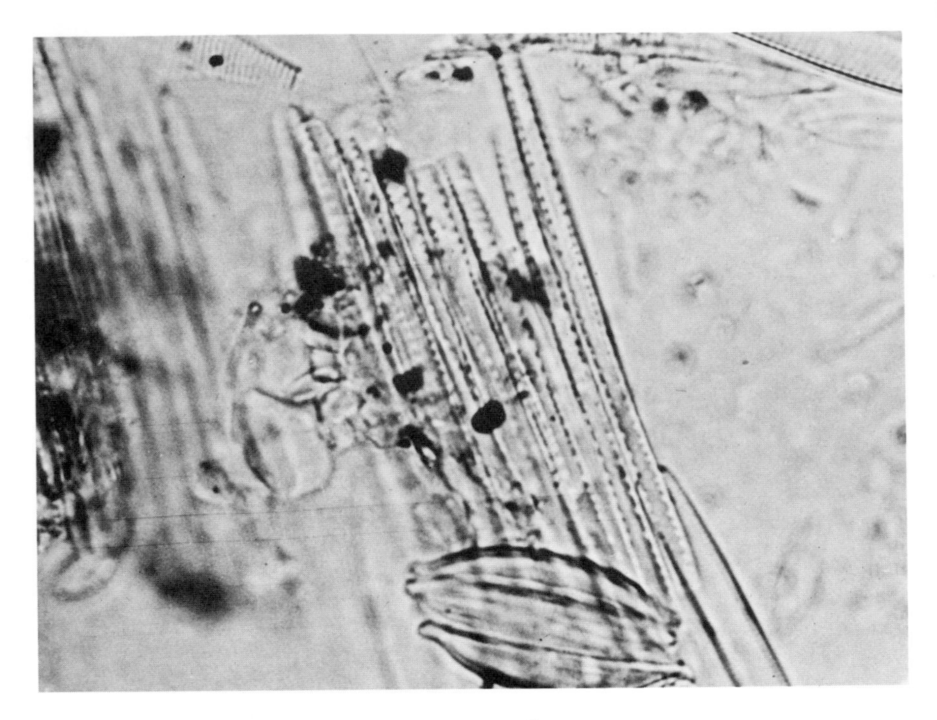

FIG. VII-17f

Observation

Look for growth of algae. Reinoculate onto agar medium to purify or use other techniques to prepare unialgal cultures. [Reference: R. C. Starr, *Am. J. Botany* **51**, 1013-44 (1964).]

130. Isolation of Algae Using Potassium Tellurite as a Bacteriostatic Agent

Materials

Potassium tellurite	
Marine medium	
Agar	1.0 g
Potassium tellurite	0.01 g
Soil extract (optional)	0.05 ml
Seawater to	100 ml
Freshwater medium	
Soil extract (optional)	0.05 ml

FIG. VII-17g

Agar	1.0 g
Potassium tellurite	0.01 g
Knop's (or other algal medium) to	100 ml

Procedure

1. Prepare agar plates with either the marine or freshwater medium and dry surface by incubating at 37°C for 24 hours.
2. Streak surface with liquid material from liquid portion or aerobic mud layer of Winogradsky column or other source of algae.
3. Incubate plates inverted for 1-3 weeks in light at 20°-30°C.

Observation

Algae will grow on surface of nutrient agar and the tellurite will inhibit the growth of contaminating bacteria permitting the algae to grow away from their contaminants. The hopefully uncontaminated algae are then transferred to fresh sterile medium. [Reference: S. C. Ducker and L. G. Willoughby, *Nature* **202**, 210 (1964).]

131. ISOLATION OF ALGAL CULTURES WITH CAFFEINE

Materials

Mineral medium for algae
Caffeine 0.01-0.03 *M*

Procedure

1. Inoculate liquid or solid mineral-medium containing caffeine (0.01-0.03 *M*), with soil or water sample containing desired algae.

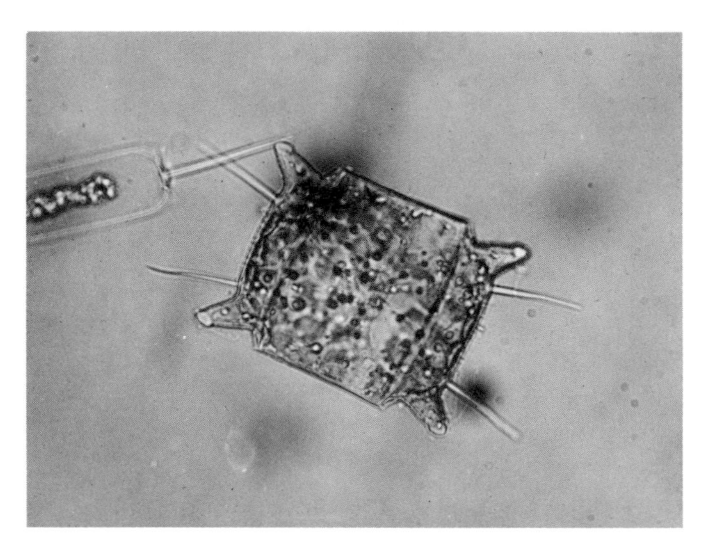

Fig. VII-17h

Fig. VII-17i

2. Incubate in light until algae are grown. Avoid media containing nucleic acid derivatives; these may annul caffeine.

Observation

Caffeine often inhibits growth of fungi and protozoa while algae are unharmed; their multiplication may even be stimulated. [Reference: S. W. Bowne, Jr., *Nature* **204**, 801 (1964).]

132. Enrichment for Freshwater Algae (Especially Blue-Green Algae)

Materials

250-ml flasks

Mineral medium A*

$MgSO_4 \cdot 7\ H_2O$	0.025 g
KH_2PO_4	0.1 g
$Ca(NO_3)_2 \cdot 4\ H_2O$	0.0025 g
KNO_3	0.005 g
Ferric citrate	0.1 g
Water to	100 ml

Mineral medium B

$MgSO_4 \cdot 7\ H_2O$	0.015 g
K_2HPO_4	0.1 g
$Ca(NO_3)_2 \cdot 4\ H_2O$	1.0 g
$NaNO_3$	0.1 g
EDTA	0.5 mg
$Fe_2(SO_4)_3 \cdot 6\ H_2O$	0.4 mg
H_5 Microelements*	1.0 ml
Water to	100 ml
pH	7.6

Agar 0.1–1.5 g/100 ml of either medium A or B may be added to solidify them.

Procedure

1. Add 50–100 ml of either medium A + $NaCO_3$ or medium B to a flask which is then inoculated with 1.0 g soil or 1.0 ml water.

*Na_2CO_3 0.15 g/100 ml is autoclaved separately and added aseptically to medium A when both are cool to avoid precipitation.

**H_5 Microelements Stock Solution:* $ZnSO_4 \cdot 7\ H_2O$, 0.88 g; $MnCl_2 \cdot 4\ H_2O$, 0.14 g; MoO_2, 0.07 g; $CuSO_4 \cdot 5\ H_2O$, 0.16 g; $Co(NO_3)_2 \cdot 6\ H_2O$, 0.05 g; water to 100 ml.

The top liquid or soil from a freshwater or terrestrial Winogradsky column is an excellent source of algae.

2. Flasks are incubated at 20°–30°C in the light until algal growth appears. Incubation in fluorescent light tends to select for blue-green algae.
3. Single species algal colonies may be isolated by streaking material from the enrichment flasks onto the appropriate medium A or B solidified with 1–1.2% agar.

Observation

Algae appear as colored growth on the top, sides, or throughout the liquid in the flask. They may form individual colonies or streaks of growth on agar. Some are motile. Colors may vary depending on the species. Identification may be made by wet mount examination under the microscope (400–1000 ×). Blue-green algae are conspicuous in that like bacteria they lack any visible internal structure (nucleus, chloroplasts, mitochondria, vacuoles) when examined by ordinary microscopy. [Reference: Modified from W. A. Kratz, and J. Myers, *Am. J. Botany* **42**, 282–7 (1955).]

133. Isolation of Marine Diatoms

Materials

Millipore filter discs with an average pore size of 5 μ.
Millipore filter
Nutrient medium

Sodium acetate · 3 H_2O	0.01 g
Yeast extract	0.05 g
Trypticase	0.05 g
Peptone	0.05 g
Agar	1.0 g
Aged seawater to	100 ml
pH after autoclaving	7.6–7.8

Procedure

1. Sterilize nutrient agar medium and prepare agar plates. Filter 100 ml of seawater, marine mud suspension, plankton tow, or fluid from top of Winogradsky column through filter disc.
2. Place disc right side-up on surface of nutrient agar plate and incubate at 22°–25°C in the light.

Observation

Examine plates periodically under binocular dissecting microscope for diatom tracts radiating out from the disc. Bacterial and algal colonies tend to remain near disc. Remove agar block containing diatoms from tract as far as possible from disc with a sterile loop or needle and inoculate (algal side down) onto surface of nutrient medium, incubate, and examine for new diatom tracts radiating from agar block. (See Fig. VII-17.) [Reference: T. J. Starr, *Texas Rept. Biol. Med.* **17**, 624–5 (1959).]

134. Enrichment and Isolation of Marine Diatoms

Materials

125-ml flask
Petri dishes
Nutrient medium

$Ca(NO_3)_2 \cdot 4\ H_2O$	0.01 g
K_2HPO_4	0.002 g
$Na_2SiO_3 \cdot 9\ H_2O$	0.005 g
Fe (as $Fe(NH_4)_2(SO_4)_2 \cdot 6\ H_2O$)	0.05 mg
Zn (as $ZnSO_4 \cdot 7\ H_2O$)	0.03 mg
B (as H_3BO_3)	0.01 mg
Co (as $CoSO_4 \cdot 7\ H_2O$)	0.01 mg
Cu (as $CuSO_4$ anhydrous)	0.01 mg
Mn (as $MnSO_4 \cdot H_2O$)	0.01 mg
Mo (as $(NH_4)_6Mo_7O_{24} \cdot 4\ H_2O$)	0.01 mg
Vitamin B_{12}	0.1 μg
Tryptone	0.1 g
Filtered seawater to	100 ml

Add agar 1.0 g/100 ml to make nutrient medium into solid medium.

Procedure

1. Sterilize flasks (cotton or screw-topped) with 50 ml of nutrient medium.
2. Inoculate with 1.0 ml of fresh seawater or 1.0 g of marine mud. A good source is the liquid or mud from the top of a marine Winogradsky column.
3. Incubate at 20°–30°C in the light for 1–3 weeks.

Observation

Look for golden brown or yellow diatoms growing at surface or glass-liquid interface. Diatoms may be isolated by streaking material from the flask onto the surface of nutrient agar medium in a Petri dish which is incubated in light at 20°-30°C for 1-3 weeks. Prevent evaporation by sealing plates with parafilm or placing them in plastic sandwich bag. Examine yellow or golden-brown colonies for diatoms by making a wet mount and examining it under microscope (100-400 ×). (See Fig. VII-17.) [Reference: J. C. Lewin, and R. A. Lewin, *Can. J. Microbiol.* **6**, 127-34 (1960).]

135. ISOLATION OF *Prymnesium parvum*, A CHRYSOMONAD EXCRETING AN ICHTHYOTOXIN

Materials

Nutrient medium

NaCl	0.1 g	$ZnSO_4 \cdot 7\,H_2O$	1.5 μg
$MgSO_4 \cdot 7\,H_2O$	0.03 g	Tris (hydroxy-methyl)amino-methane	10.0 mg
KCl	8.0 mg		
$CaCl_2 \cdot 2\,H_2O$	1.0 mg		
H_3BO_3	0.1 mg	Thiamine · HCl	0.1 μg
Na_2HPO_4	0.5 mg	Ethionine	0.1 g
$MnCl \cdot 4\,H_2O$	0.05 mg	Vitamin B_{12}	1.0 μg
$FeCl_2 \cdot 6\,H_2O$	0.01 mg	Water to	100 ml
$NaMoO_4 \cdot 4\,H_2O$	0.01 mg	pH	8.2-8.4

Procedure

1. Sterilize nutrient medium and inoculate tubes containing 10 ml of sterile medium with soil or water.
2. Incubate in the light at 25°C for 4-5 days.
3. Reinoculate into fresh media every 4-5 days. *Prymnesium parvum* is unusual in being able to use ethionine as the sole source of nitrogen. Ethionine is usually toxic at this concentration.

Observation

Look for motile brown algae. Reinoculate into fresh media until cultures are unialgal. Test supernatant of unialgal cultures for ichthyotoxin by exposing fish to supernatant and looking for death of fish. Toxin is light sensitive. [Reference: M. Rahat, and I. Dor, *Israel J. Botany* **16**, 42-4 (1967).]

136. Isolation of the Snow Alga, *Chromulina*

Materials

Gray, olive green, or pink patches of snow

Nutrient medium BMT

Part A	
$Ca(NO_3)_2 \cdot 4\ H_2O$	0.4 g
Water to	50.0 ml
Part B	
KNO_3	0.1 g
KH_2PO_4	0.1 g
$MgSO_4 \cdot 7\ H_2O$	0.1 g
Water to	50.0 ml

Combine Parts A and B in equal quantities. Add trace element solution* 0.1 ml/100 and Tris-HCl† 0.25%, pH 4.0.

Nutrient medium—Tris

KNO_3	1.0 g
KH_2PO_4	0.1 g
$MgSO_4 \cdot 7\ H_2O$	0.1 g
Trace element solution*	0.1 ml
Tris-HCl†	1.0 g
Soil-water medium (see Method No. 129)	5.0 ml
Water to	100 ml
pH	4.5

Procedure

1. Place snow material in 100 ml of the acid media (above) in a 250-ml flask.
2. Incubate at 5°C in 100–300 ft-c of white light for several weeks. Low temperature necessary.

**Trace element solution*: H_3BO_3, 1.0 g; $CuSO_4 \cdot 5\ H_2O$, 0.15 g; EDTA (Versene) (ethylenediaminetetraacetic acid), 5.0 g; $ZnSO_4 \cdot 7\ H_2O$, 2.2 g; $CaCl_2$, 0.5 g; $MnCl_2 \cdot 4\ H_2O$, 0.5 g; $FeSO_4 \cdot 7\ H_2O$, 0.5 g; $CoCl_2 \cdot 6\ H_2O$, 0.15 g; $(NH_4)_6Mo_7O_{24} \cdot 4\ H_2O$, 0.10 g; Water to 100 ml; pH 6.5.

†Tris-HCl (Trizma-HCl) is neutralized hydrochloride of tris(hydroxymethyl)aminomethane, Sigma Chemical Co.

Observation

Look for ovoid-elliptical cells 5-12 μ long × 5-6 μ wide × 2-3 μ in diameter with single flagellum and one bilobed yellow-brown chloroplast; cells die quickly in wet mounts from heat of illumination. [Reference: J. R. Stein, *Can. J. Botany* **41**, 1367-70 (1963).]

V. Free-Living Protozoa

137. Media for Cultivating Protozoa Nonaxenically,

Materials

Flagellates With Chromatophores

(a)	Peptone (or tryptone)	0.2 g
	KH_2PO_4	0.025 g
	$MgSO_4 \cdot 7\ H_2O$	0.025 g
	KCl	0.025 g
	$FeCl_3 \cdot 6\ H_2O$	trace
	Sodium acetate · 3 H_2O	0.2 g
	Water to	100 ml
(b)	Peptone (or tryptone)	0.25 g
	KNO_3	0.05 g
	KH_2PO_4	0.05 g
	$MgSO_4 \cdot 7\ H_2O$	0.01 g
	NaCl	0.01 g
	Sodium acetate · 3 H_2O	0.25 g
	Glucose	0.2 g
	Water to	100 ml

Phagotrophic Flagellates Without Chromatophores

(a)	Milk	1.0 ml
	Soil extract	10.0 ml
	Water to	100 ml

Flagellates Without Chromatophores

(a) 4 rice grains in 150 ml water in a finger bowl. Let dish stand for a few days, then inoculate with flagellate material.

Amebae and Ciliates

(a) 4 rice grains in 150 ml water in a finger bowl. Let dish stand for a few days then inoculate with protozoan material.

(b) Hay-infusion. Boil 3-4 2-inch pieces of hay in 100 ml water and allow to cool. Let infusion stand for several days then inoculate with protozoan material.

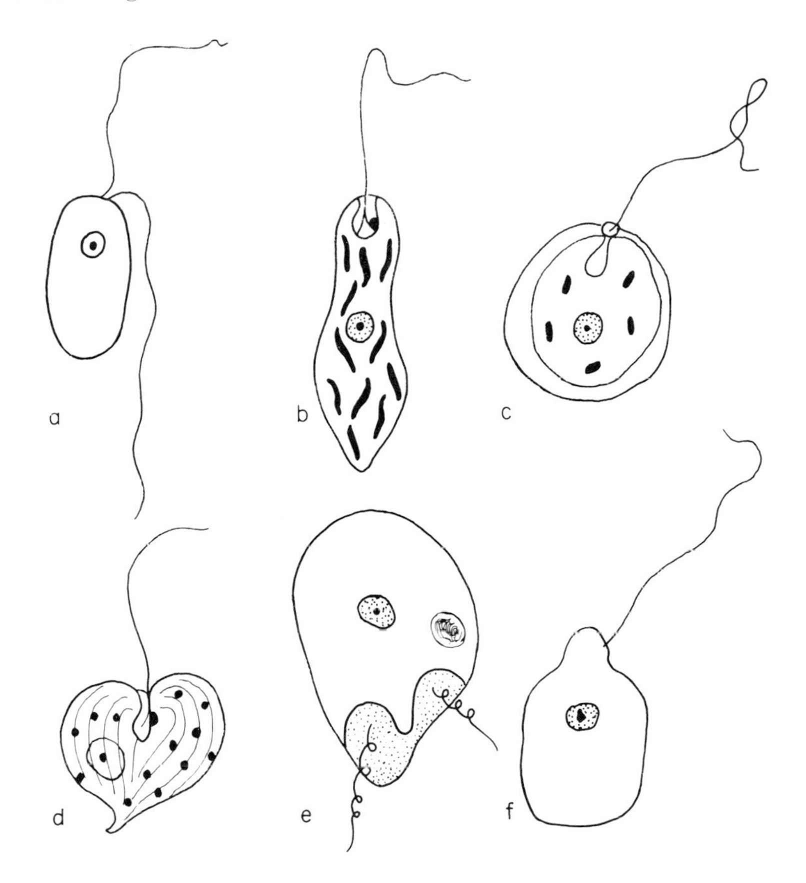

FIG. VII-18. Some common freshwater and marine flagellates. (a). *Bodo* (10-15 μ), freshwater; (b). *Euglena* (80-150 μ), freshwater; (c). *Trachelomonas* (10-20 μ), freshwater; (d). *Phacus* (40-100 μ), freshwater; (e). *Oxyrrhis marina* (20-30 μ), marine; (f) *Oikomonas* (5-20 μ), marine.

(c) Lettuce-infusion. Boil 2 square inch pieces of lettuce in 100 ml water and cool. Let infusion stand for several days then inoculate with protozoan material.

These may be supplemented with a pinhead amount of hard boiled egg yolk before adding protozoan material. [Reference: R. R. Kudo, "Protozoology." Thomas, Springfield, Illinois, 1966.]

138. ENRICHMENT FOR FREE-LIVING FLAGELLATES AND OTHER PROTOZOA. See method No. 128-129.

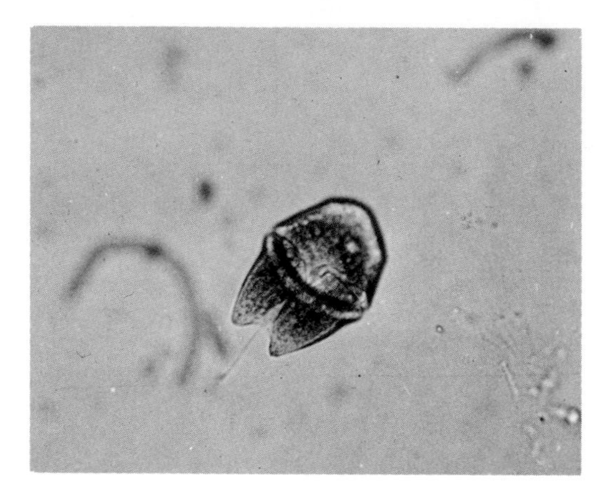

FIG. VII-19a

FIG. VII-19. Some marine dinoflagellates. (a). *Gyrodinium* sp.; (b). *Ceratium tripos* (from Dr. J. J. Lee).

139. CONCENTRATING PHOTOTACTIC FLAGELLATES

Materials

500-ml round flask with flat bottom and radially directed side arm.
Marble to fit top of side arm.

Procedure

1. Fill flask with fluid containing flagellates and place in window so that side arm is at right angles to the window and projecting into room (See Fig. VII-20). Retain in this position for 30–60 minutes or as long as is necessary for the flagellates to collect in side arm.
2. Place side arm vertically so that marble closes side arm tube.
3. Pour off excess fluid. Flagellates should be concentrated in side arm tube.
4. Procedure may be modified so that only side arm is in light by placing flask in a closed box with side arm projecting out (See Fig. VII-20.) [Reference: J. D. Meeuse, *Arch. Mikrobiol.* **45**, 423–24 (1963).]

FIG. VII-19b

140. ISOLATION OF AXENIC AMEBAE BY MIGRATION ON AGAR SURFACE

Materials

BM medium(add agar, 1.5%, for nonnutrient agar)

NaCl	12.0 mg
$MgCl_2 \cdot 6\ H_2O$	0.3 mg
$FeSO_4 \cdot 7\ H_2O$	0.3 mg

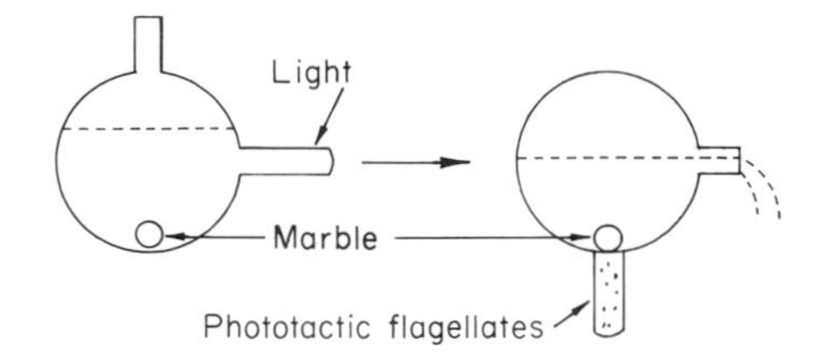

FIG. VII-20. Position for concentrating phototactic flagellates.

Na_2HPO_4(0.1 *M*)	1.0 ml
KH_2PO_4(0.1 *M*)	1.0 ml
Water to	100 ml
pH	6.0-8.0

Soil
Bacteria or yeast (washed twice in distilled water) and autoclaved.

Procedure

1. Prepare sterile BM agar plates (1.5 g agar/100 ml water).
2. Streak surface with dead bacteria or yeast.
3. Sprinkle plate with soil.
4. Incubate at room temperature for 1-3 weeks.
5. Seal plate with plastic to prevent drying.

Observation

Examine plates under dissecting microscope for amebae, *migrating away* from clumps of autoclaved cells. Transfer amebae or cysts to fresh nonnutrient agar with sterile micropipette (or Pasteur pipette). Streak as before with autoclaved cells and reisolate. Amebae may be maintained in liquid BM supplemented with autoclaved cells. Test cultures for presence of contaminants by streaking amebae on several types of nutrient media for the growth of bacteria. [Reference: R. J. Neff, *J. Protozool.* **4**, 176-8 (1957); *ibid.* **5**, 226-31 (1958).]

141. ISOLATION OF PROTOSTELIDS, MYCETOZOA (MYXOMYCETES) – Olive. See Method No. 122.

142. ISOLATION OF SLIME MOLDS – Raper. See Method No. 123.

143. SELECTED MEDIA FOR THE CULTURE OF PROTOZOA

[See catalog of laboratory strains of free-living and parasitic protozoa for additional procedures. *J. Protozool.* **5**, 1-38 (1958).]

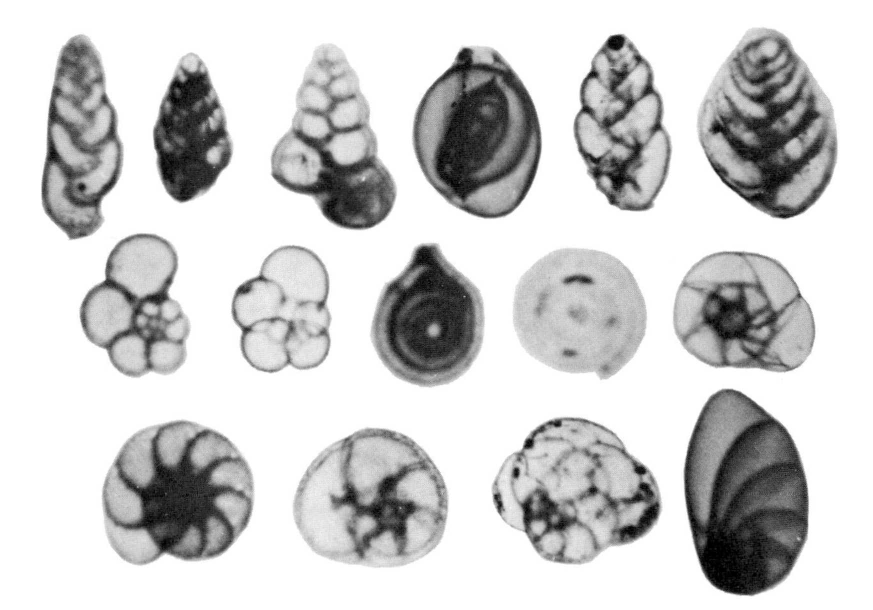

FIG. VII-21. Some common marine foraminifera and amebae. (a). Some foraminiferan skeletons; (b). *Allegromia laticolaris* and attached bacteria; (c). *Globigerinoides rubra*; (d). *Globigurina bulloides*, a planktonic foraminiferan; (e) A foraminiferan skeleton, *Protelphidium tisburense*, and diatom skeletons; (f). Flabellid ameba, *Vexillaria* sp. (From Dr. J. J. Lee.)

Materials

(*a*) For phytoflagellates

1. $NaNO_3$	10.0 mg
K_2HPO_4	2.0 mg
Soil extract	5-10 ml
Water (or seawater) to	100 ml
pH	8.0-8.2

Incubate at 14-18°C with moderate illumination.

2. Liver infusion (Oxo, Ltd., England)	10.0 mg
Tryptone (Difco, United States)	10.0 mg
KNO_3	10.0 mg
K_2HPO_4	1.0 mg
Soil extract	4-7 ml
Seawater (¼-⅔ strength) to	100 ml
pH	7.2-7.6

Incubate below 25°C with moderate illumination.

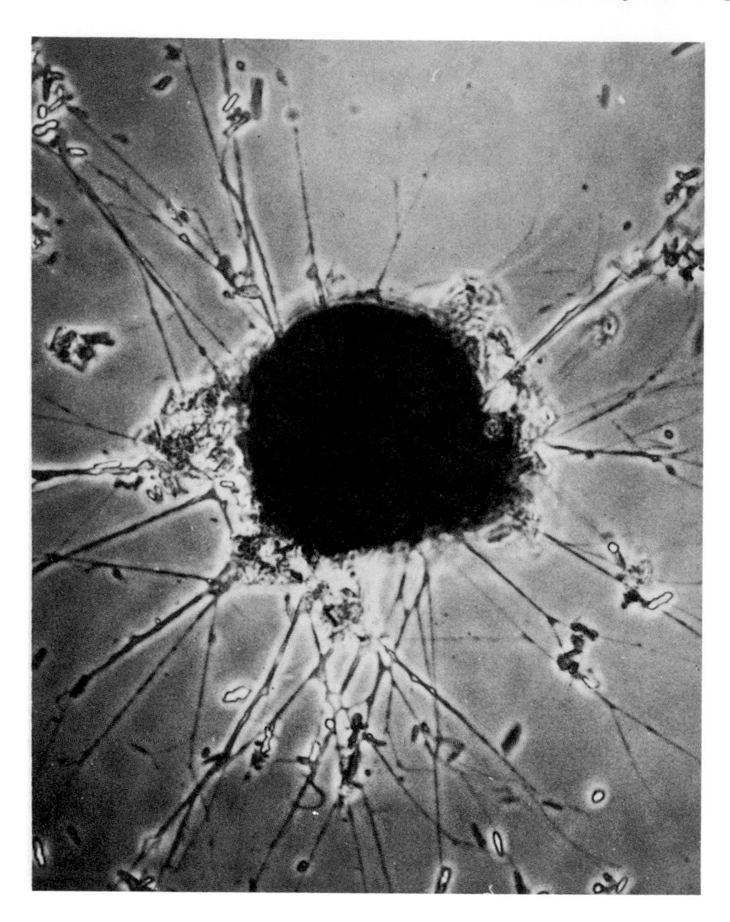

FIG. VII-21b

3. Proteose peptone (Difco)	0.1 g
KNO_3	20.0 mg
K_2HPO_4	2.0 mg
$MgSO_4 \cdot 7H_2O$	2.0 mg
Agar	1.0 g
Water to	100 ml
pH	6.0-7.0

Agar surface should be wet. Incubate at 25°C with moderate illumination.

4. $NaNO_3$	10.0 mg
K_2HPO_4	2.0 mg

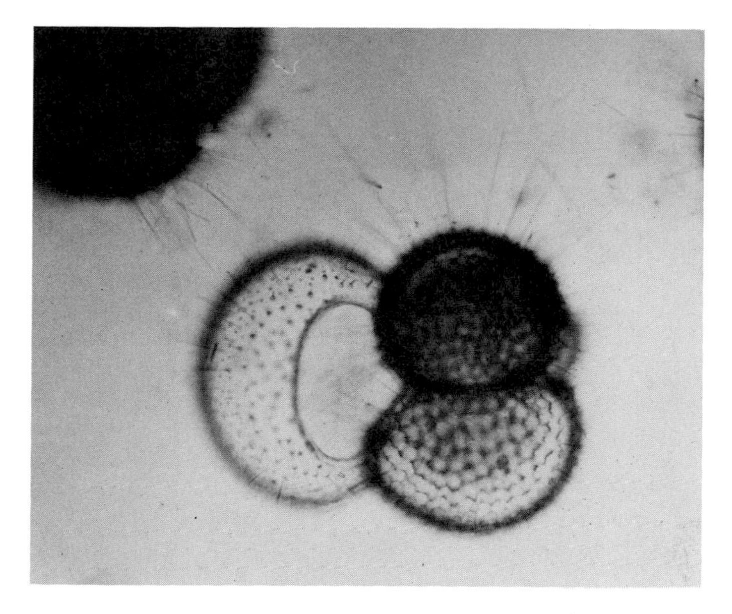

FIG. VII-21c

Soil extract	10.0 ml
NaCl	3.0 g
Seawater to	100 ml
pH	8.0–8.2

Incubate at 14°–18°C with moderate illumination.

(*b*) For mycetozoa (slime molds)

Peptone (Difco)	0.5 g
Glucose	0.5 g
Agar	2.0 g
Water to	100 ml

Inoculate slant or Petri dish containing above medium with bacteria (e.g., *Escherichia coli*) and slime mold. Incubate at 16°–25°C.

(*c*) For ameba

Peptone	2.0 g
Skim-milk powder	0.02 g
Water to	100 ml

Incubate at 24°–30°C.

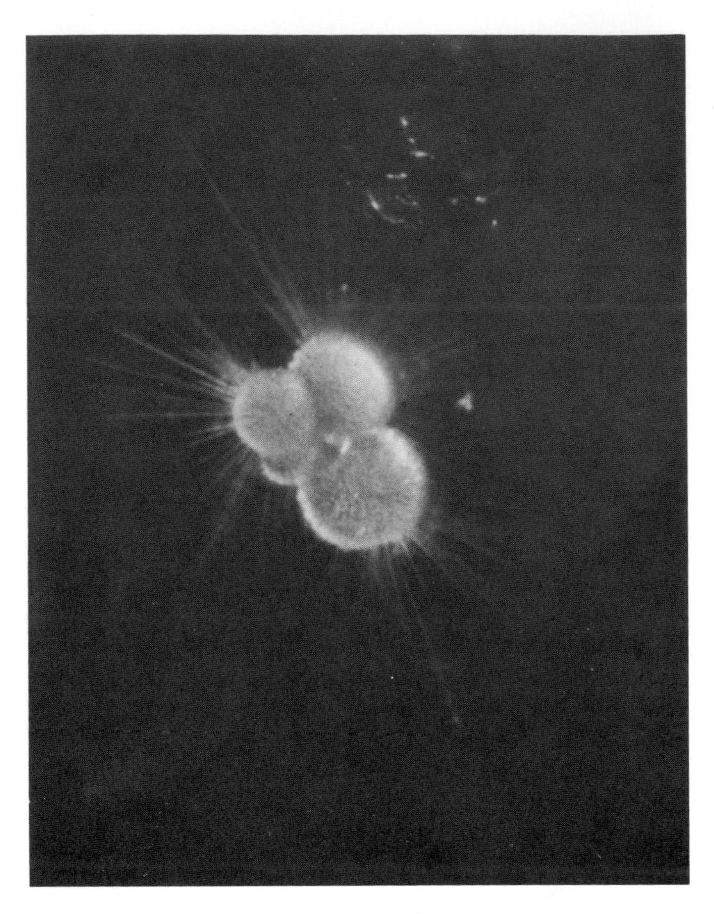

FIG. VII-21d

(*d*) For foraminifera
Diatom suspension
Seawater (salinity 3.35%) 100 ml

Incubate in subdued light at 18°-24°C.

(*e*) Ciliates
1. Dried lettuce (or Cerophyl; Cerophyl Labs., Kansas City, Missouri) 0.15 g
$CaCO_3$ 0.001 g

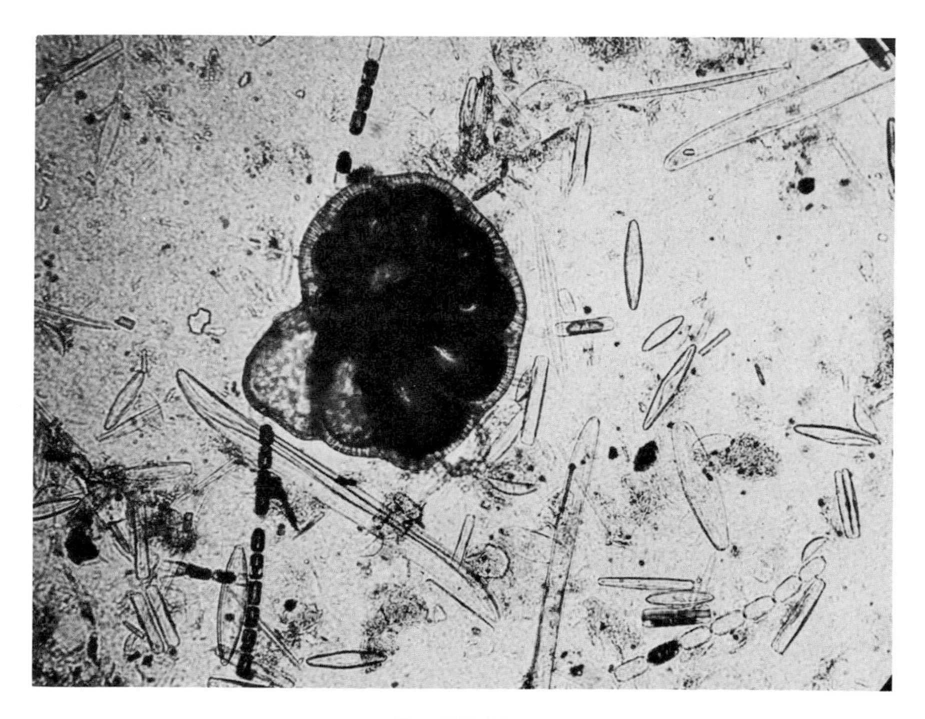

FIG. VII-21e

Water to	100 ml
pII	7.0

Heat to boil, cool, and filter. Place 5 ml in a tube, sterilize, and inoculate with bacterium (e.g., *Aerobacter aerogenes*) and ciliate. Incubate at 10°-25°C.

2. Yeast autolyzate	0.5 g
Casein*	0.25 g
K_3PO_4	0.1 g
Glucose (*add aseptically*)	0.5 g
Water to	100 ml
pH	7.0

Incubate at 10°-25°C.

*Casein solution (5%) is prepared by adding casein to hot water and adding KOH dropwise, with stirring, until casein is in solution.

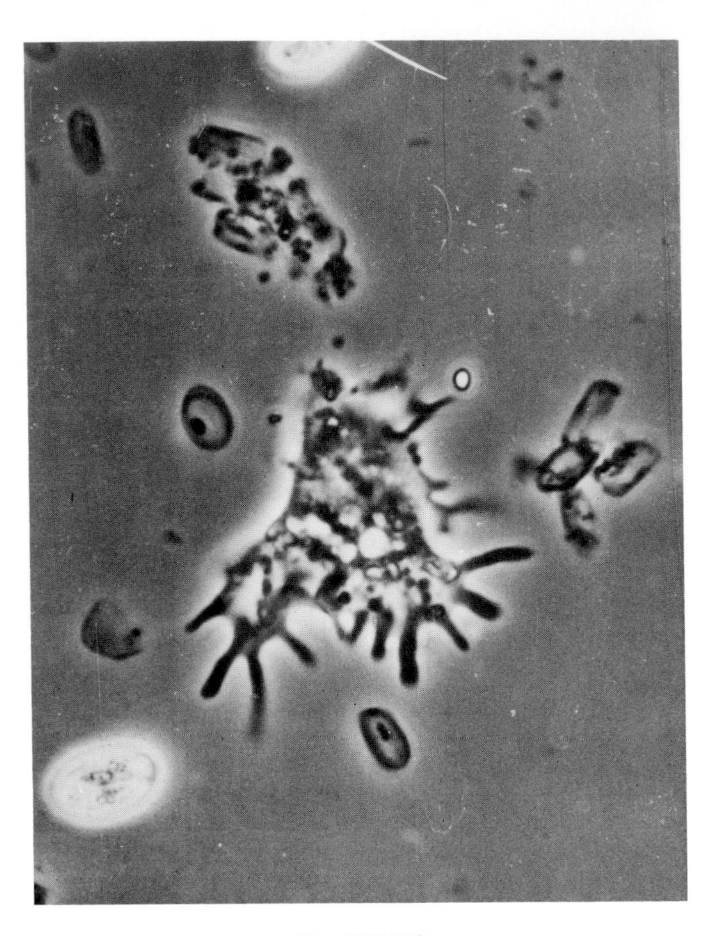

FIG. VII-21f

VI. Miscellaneous

144. Demonstration of Microorganisms on Sand Grains

Materials

Osmic Acid
Bouin's fixative (see p. 43)
Carbolfuchsin (mix solutions A and B)
Solution A

Basic fuchsin	0.3 g
Ethyl alcohol (95%)	10.0 ml

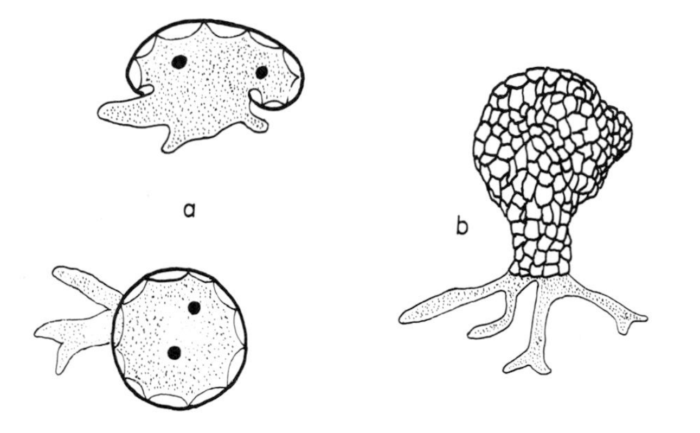

FIG. VII-22. Some common freshwater amebae. (a). *Arcella* sp. (30-300 μ); (b). *Difflugia* sp. (200-300 μ).

Solution B	
Phenol	5.0 g
Water	95 ml

Procedure

1. Treat freshwater or marine sand grains with either osmic acid vapor or Bouin's fixative.
2. Wash off fixative.
3. Stain sand grains with carbolfuchsin.

Observation

Examine sand grains under microscope (500-1000 ×). Small colonies (10-100 organisms) may be found in cracks and crevices of sand grain. Bacteria, blue-green algae, yeasts, and the early stages of brown algae have been identified. Surface sands show greatest incidence of microorganisms. [Reference: P. S. Meadows, and J. G. Anderson, *Nature* **212**, 1059-60 (1966).]

145. ENRICHMENT FOR ACID-LOVING MICROORGANISMS (BACTERIA, YEASTS, AND PROTOZOA)

Materials

Enrichment medium	
$(NH_4)_2SO_4$	0.3 g
KCl	0.01 g
K_2HPO_4	0.05 g

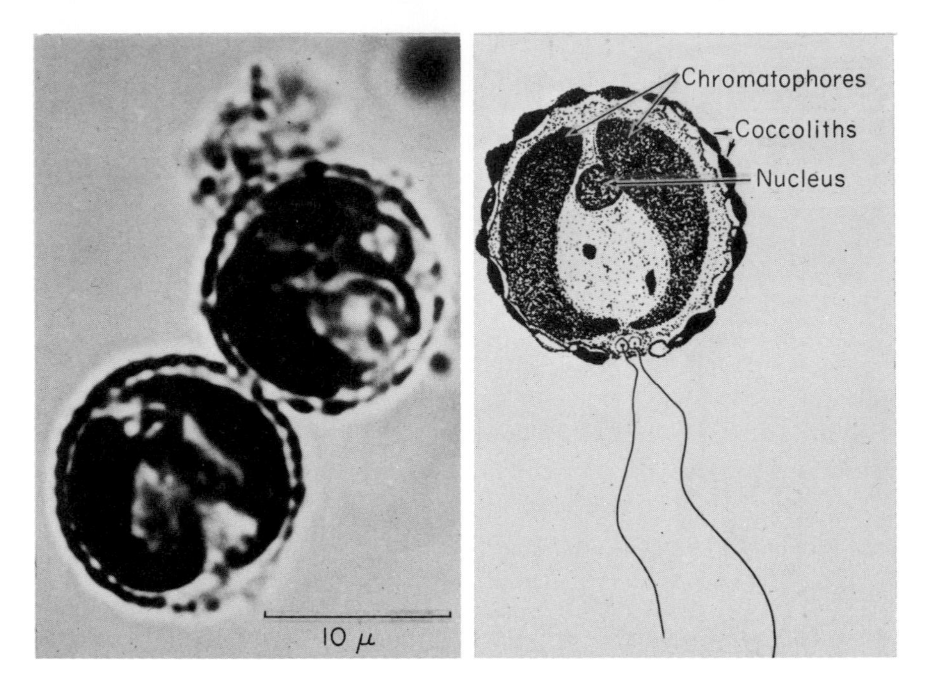

FIG. VII-23a

FIG. VII-23. Some marine and freshwater protozoa. (a). *Hymenomonas* sp., a coccolithophorid. This organism has calcareous buttons, discs or rings in an outer membrane and is common in the sea. Chrysophyta (from Dr. H. D. Isenberg); (b). *Noctiluca* sp., a common marine luminous dinoflagellate, Chrysophytya (from Dr. K. Gold). (c). *Tintinnopsis* sp., a marine ciliate (from Dr. K. Gold); (d). *Didinium* eating *Paramecium*; freshwater ciliates, stained.

$MgSO_4 \cdot 7\ H_2O$	0.05 g
$Ca(NO_3)_2 \cdot 4\ H_2O$	1.0 mg
$FeSO_4 \cdot 7\ H_2O$*	30 ml
H_2SO_4 (10 *N*)	1.0 ml
Water to	100 ml

Acid mine or bog water

Procedure

1. Innoculate 5.0 ml of acid water sample into 30 ml of enrichment medium in a 125 ml flask.
2. Incubate at 25°C in dark.

*Use 30 ml of a 14.78% solution. Solution and medium good for 2 weeks.

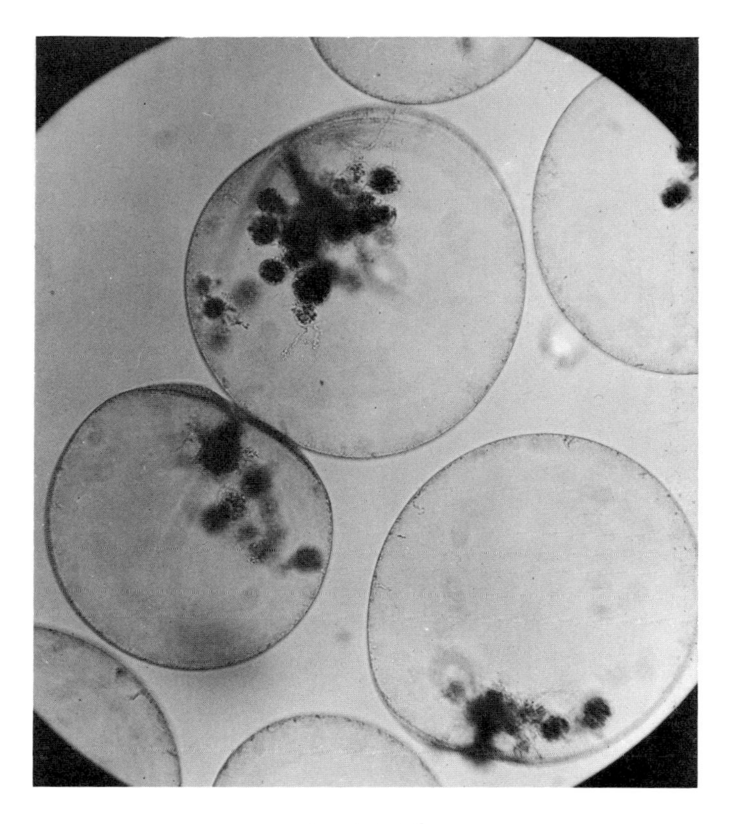

FIG. VII-23b

Observation

After 3-4 days bacteria will reach maximal population. These are likely to belong to *Ferrobacillus-Thiobacillus* group. Yeasts may also be present at this time. After 2 weeks flagellate and ameboid protozoa appear feeding on the bacteria and yeast. The flagellates are likely to belong to the order Euglenoidina. [Reference: H. L. Ehrlich, *J. Bacteriol.* **86**, 350-352 (1961).]

146. DIRECT OBSERVATION OF MICROORGANISMS GROWING ON SLIDES—ROSSI-CHOLODNY TECHNIQUE

Materials

Microscope slides
Stains for bacteria, fungi, algae

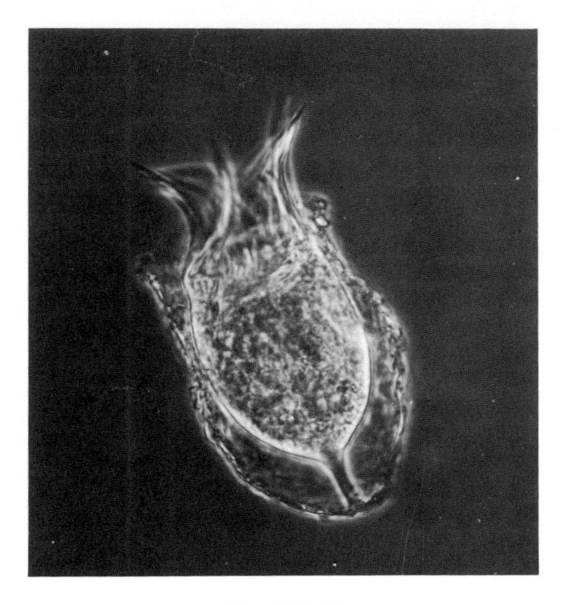

FIG. VII-23c

Nutrient agar; nutrient gelatin or agar containing any nutrient desired

Procedure

1. Clean slides may be used directly or they may be dipped in appropriate nutrient gelatin or agar and dried.
2. Bury slides in soil or mud or place slides on submerged floats in water. Place appropriate marker rocks or other marker over soil and a buoy over water sampler. Allow slides to remain in soil, mud, or water for 1–3 months.
3. Remove and examine microscopically by wet mount or slides may be dried, fixed, and stained by procedures described elsewhere in this book.

Observation

Microscopically observe colonies of bacteria, fungal mycelia, algae, or protozoa present.

[Reference: (1) G. M. Rossi, *Ital. Agr.* 4 (1928); (2) N. G. Cholodny, *Arch. Mikrobiol.* **1**, 620 (1930).]

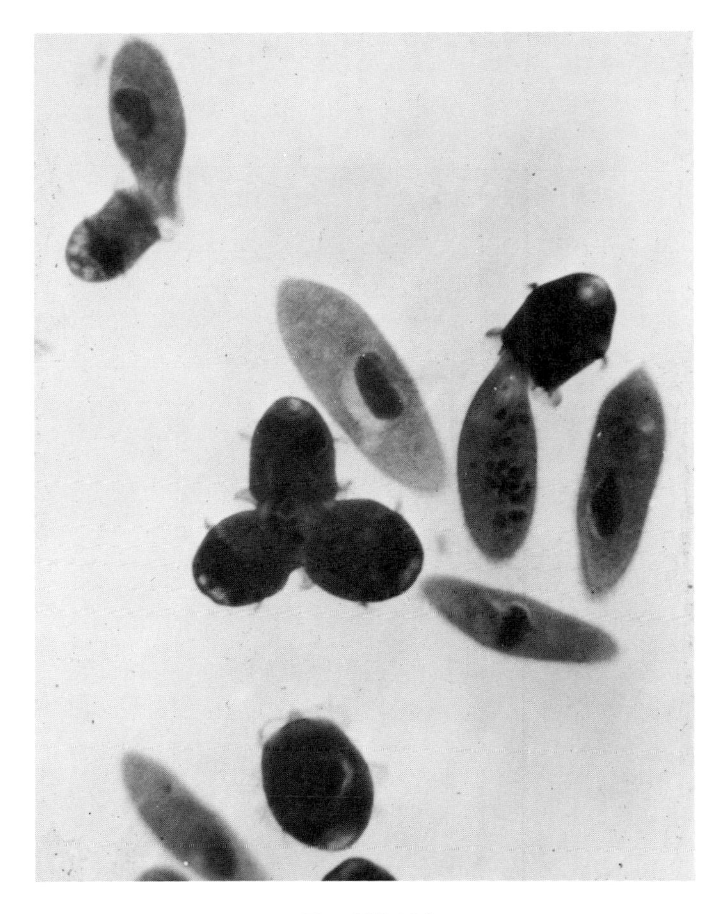

Fig. VII-23d

SELECTED REFERENCES

"Advances in Applied Microbiology." Academic Press, New York.

Ainsworth, G. C. and Sussman, A. S. (eds.), "The Fungi," Vols. I-III. Academic Press, New York, 1965-1967.

Alexander, M. "Introduction to Soil Microbiology." Wiley, New York, 1961.

Alexander, M. Biochemical ecology of soil microorganisms. *Ann. Rev. Microbiol.* **18**, 217-252 (1964).

Alexopoulos, C. J. "Introductory Mycology." Wiley, New York, 1962.

Alexopoulos, C. J. The myxomycetes II. *Botan. Rev.* **29**, 1-78 (1963).

Alexopoulos, C. J. and Bold, H. C. "Algae and Fungi." Macmillan, New York, 1967.

"Annual Review of Microbiology." Annual Reviews, Inc., Palo Alto, California.

Boney, A. D. "A Biology of Marine Algae." Hutchinson Educational Ltd., London, 1966.

Bonner, J. T. "The Cellular Slime Molds." Princeton University Press, Princeton, New Jersey, 1967.

Brock, T. D. Microbial ecology and applied microbiology. *Advan. Appl. Microbiol.* **8**, 61-75 (1966).

Brock, T. D. "Principles of Microbial Ecology." Prentice-Hall, Englewood Cliffs, New Jersey, 1966.

Burges, A. "Microorganisms in the Soil." Hutchinson & Co., Ltd., London, 1958.

Burges, A. and Raw, F. (eds.). "Soil Biology." Academic Press, London and New York, 1967.

Chapman, V. J. "The Algae." Macmillan, London, 1962.

Cochrane, V. W. "The Physiology of Fungi," Wiley, New York, 1958.

Dawson, E. Y. "How to Know the Seaweeds." W. C. Brown, Dubuque, Iowa, 1956.

Dawson, E. Y. "Marine Botany," Holt, Rinehard and Winston, New York, 1966.

Fogg, G. E. "Algal Cultures and Phytoplankton Ecology," University of Wisconsin Press, Madison, Wisconsin, 1965.

Fritsch, F. E. "The Structure and Reproduction of the Algae," Vols. I and II. Cambridge University Press, London and New York, 1948, 1959.

Frobisher, M. "Fundamentals of Microbiology." Saunders, Philadelphia, Pennsylvania, 1962.

Gibbs, B. M. and Shapton, D. A. "Identification Methods for Microbiologists." Academic Press, London and New York, 1967-1968.

Gurr, E. "The Rational Use of Dyes in Biology." Williams & Wilkins, Baltimore, Maryland, 1965.

Harvey, H. W. "The Chemistry and Fertility of Seawaters." Cambridge University Press, London and New York, 1957.

Hawker, L. E., Linton, A. H., Folkes, B. F. and Carlile, M. J. "An Introduction to the Biology of Microorganisms." St. Martin's Press, New York, 1960.

Hawker, L. E. "Fungi." Hutchinson & Co., Ltd., London, 1966.

Jahn, T. L. "How to Know the Protozoa," W. C. Brown, Dubuque, Iowa, 1949.

Kriss, A. E. "Marine Microbiology (Deep Sea)." Wiley (Interscience), New York, 1963.

Kudo, R. R. "Protozoology." Thomas, Springfield, Illinois, 1966.
Lamanna, C. and Mallette, M. F. "Basic Bacteriology." Williams & Wilkins, Baltimore, Maryland, 1965.
Lewin, R. A. (ed.). "Physiology and Biochemistry of Algae." Academic Press, New York, 1962.
Microbial Ecology. *7th Symp., Soc. Gen. Microbiol. 1957.* Cambridge University Press, London and New York.
Moore, H. H. "Marine Ecology." Wiley, New York, 1958.
Morris, Ian. "An Introduction to the Algae." Hutchinson & Co., Ltd., London, 1967.
Oppenheimer, C. H. (ed.). "Symposium on Marine Microbiology." Thomas, Springfield, Illinois, 1963.
Parkinson, D. and Waid, J. S. "The Ecology of Soil Fungi." Liverpool University Press, Liverpool, 1960.
Phaff, H. J., Miller, M. W. and Mrak, E. M. "The Life of Yeasts," Harvard Univ. Press, Cambridge, Massachusetts, 1966.
Prescott, G. W. "How to Know the Fresh-water Algae," W. C. Brown, Dubuque, Iowa, 1964.
Prescott, G. W. "Algae of the Western Great Lakes Area." Cranbrook Institute of Science Press, Bloomfield Hills, Michigan, 1966.
Prescott, G. W. "The Algae: A Review." Houghton Mifflin Co., Boston, Massachusetts, 1968.
Round, F. E. "The Biology of the Algae." Edward Arnold Ltd., London, 1965.
Schlegel, H. G. (ed.). "Anreicherungskultur und Mutantenauslese." G. Fischer Verlag, Stuttgart, 1965.
Skerman, V. B. D. "A Guide to the Identification of the Genera of Bacteria." Williams & Wilkins, Baltimore, Maryland, 1959.
Sparrow, F. K., Jr. "Aquatic Phycomycetes." University of Michigan Press, Ann Arbor, Michigan, 1960.
Stanier, R. Y., Doudoroff, M. and Adelberg, E. A. "The Microbial World." Prentice-Hall, Englewood Cliffs, New Jersey, 1963.
Taylor, W. R. "Marine Algae of the Northeastern Coast of North America." University of Michigan Press, Ann Arbor, Michigan, 1957.
Taylor, W. R. "Marine Algae of the Eastern Tropical and Subtropical Coasts of the Americas." University of Michigan Press, Ann Arbor, Michigan, 1960.
Waksman, S. A. "The Actinomycetes, Vol. 2: Classification Identification, and Description of Genera and Species." Williams & Wilkins, Baltimore, Maryland, 1961.

Subject Index

C

D

E

F

G

H

I

W

Y

Z